Charles Seale-Hayne Library
University of Plymouth
(01752) 588 588
LibraryandITenquiries@plymouth.ac.uk

Climate and Energy
Systems

Climate and Energy Systems

A review of their interactions

Jill Jäger

Consultant Climatologist
Karlsruhe, F.R. Germany

JOHN WILEY & SONS

Chichester · New York · Brisbane · Toronto · Singapore

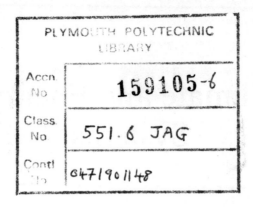
Library of Congress Cataloging in Publication Data:

Jäger, Jill.
  Climate and energy systems.
  Includes indexes.
  1. Climatic changes.   2. Power resources—Climatic factors.   3. Waste heat—Climatic factors.   I. Title.
  QC981.8.C5J33   1983   551.6   82-21925

  ISBN 0 471 90114 8

British Library Cataloguing in Publication Data:
Jäger, Jill
  Climate and energy systems.
  1. Power resources—Environmental aspects
  I. Title
  333.79   HD9502.A2

  ISBN 0 471 90114 8

Typeset in Great Britain by
Pintail Studios Ltd, Ringwood, Hampshire
Printed in Great Britain by
St Edmundsbury Press, Bury St Edmunds, Suffolk

Contents

Preface

The possibility of undesirable—and perhaps even irreversible—climate changes due to energy conversion and use could constrain the future development of energy systems. The purpose of this book is to examine what is known about this possibility.

This examination can devote most of its attention to issues of climate, owing to the fortunate fact that a globally comprehensive perspective of the world's energy future for the next 50 years already exists and can be used as the point of departure. This perspective has been prepared by the International Institute for Applied Systems Analysis, Laxenburg, Austria, and has been published (Energy Systems Program Group of the International Institute for Applied Systems Analysis, Wolf Häfele, Program Leader (1981) *Energy in a Finite World: Volume 1. Paths to a Sustainable Future; Volume 2. A Global Systems Analysis.* Cambridge, Massachusetts: Ballinger Publishing Company).

For the purpose of this book, three main points from the IIASA study are significant:

- The demand projections suggest that by the year 2030 energy demand is likely to be two to five times larger than it is now.
- Contrary to many expectations, during the next 50 years the world will have to rely on increasing amounts of fossil fuels, including some that we now think of as 'unconventional'.
- Since the earth's fossil-fuel resources are limited, the world will have to shift eventually to renewable sources of energy (such as nuclear and solar power), but this transition will take place for the most part after the year 2030.

On the basis of these findings, this book examines this question: Could the large-scale deployment and use of fossil-fuel, nuclear, and solar-energy systems cause climatic changes?

After an introductory chapter that refines this question by presenting evidence related to it, the second chapter describes briefly the physical basis of the climate

vii

system and how the influence of human activities on it can be assessed, in order to make the rest of the book accessible to readers who are not professional climatologists.

Many knowledgeable observers feel that the most serious climate issue related to the world's future energy strategy is a result of the release of CO_2 by fossil-fuel burning and its build-up in the atmosphere. Thus, the third chapter—the longest in the book—examines this issue. It concludes that, while CO_2 from burning fossil fuels could have significant influence on global climate within the foreseeable future, there are many uncertainties to be resolved before reliable predictions of the effects of particular energy strategies can be made.

The next three chapters consider the possible climatic effects of waste heat, man-made particles and gases other than CO_2, and solar energy systems, with these results:

- The release of waste heat on the scale foreseen for the next 100 years will not have a significant effect on the global climate, but regional and local effects could occur.
- For particles and gases other than CO_2 (including sulphur dioxide (SO_2), methane (CH_4), and ammonia (NH_3), the work with relatively simple models done so far does not support definitive estimates of potential climatic effects.
- There is a possibility that deploying certain renewable energy technologies on a large scale could have climatic effects by, for example, changing the earth's surface energy balance, roughness, or wetness. However, the large number of different technologies and the likelihood that they will be used in a decentralized mode suggest that climatic effects will generally be minimal.

Chapter 7 takes a view from the other direction: rather than discussing energy system effects on climate, it gives a brief overview of climatic effects on energy supply and demand.

The final chapter brings together the book's findings and conclusions.

This book was developed within the context of the IIASA global energy study as an important contribution to its work. Therefore, it was not intended to be a comprehensively detailed study of the world's climate system; rather, it takes up only the aspects of climate related to the book's central question. Similarly, since the IIASA energy study dealt with such matters, no detailed analyses of energy technologies or scenarios are reported here; rather, this book relies on the work of the IIASA energy study, summarizing here only the facts necessary for this book's purpose. Finally, this book adds little new scientific knowledge to the subjects it treats; rather, it brings together the very scattered knowledge that we have so far and focuses it on the book's central question: Could the large-scale deployment and use of fossil-fuel, nuclear, and solar-energy systems cause climatic changes?

Thus, the book's focus is on the potential scales of the interactions between two complex systems: the climate system and the world's energy system. And the emphasis is on global-scale interactions, because they are of most interest in the international and interdisciplinary context within which the book was prepared.

In this regard, I should add that the book was not intended to be exclusively for climatologists, although it will certainly be useful for those interested in learning about the interdisciplinary applications of their science. The content of the book mainly reflects the needs of energy systems analysts, environmental scientists, engineers and, perhaps, those making decisions about energy policy. During the course of the study at IIASA, I repeatedly found that such people would welcome a source of general information on the climate system and the ways in which it interacts with the energy system. In addition to general information, the book contains references to more detailed material, permitting the interested reader to undertake further research.

The work on which this book is based was mostly carried out while I was a member of the Energy Systems Program of the International Institute for Applied Systems Analysis during the period August 1976 to August 1978; however, additional effort during the winter of 1982 brought the earlier work up to date. Professor Wolf Häfele, Leader of the Energy Systems Program, gave his support and encouragement throughout the work.

The IIASA Subtask on Energy and Climate, from which this manuscript emerged, was supported by the United Nations Environmental Program.

Many people at IIASA contributed to the study's results, and many colleagues from other scientific institutions helped by sending material or discussing ideas. All of their contributions are gratefully acknowledged.

I owe a special word of thanks to Ingrid Teply-Baubinder, who gave me invaluable help, both during the IIASA project and during the subsequent preparation of this book.

Jill Jäger
Karlsruhe

CHAPTER 1

Introduction

It has already been observed that human activities can have an influence on the climate. For example, cloudiness and precipitation are altered over and for several kilometres downstream of urban-industrial areas as a result of the heat, gases, and particles they add to the atmosphere. Similarly, there are well documented differences between urban and rural climates, including the 'urban heat island' effect due to changes in the characteristics of the earth's surface and heat additions in urban areas. Up to now, the changes in climate due to human activities appear to have been mostly on the local scale, although regional occurrences of air pollution suggest that larger-scale effects are possible.

In recent years mankind has become more aware of the sensitivity of the climate system. This awareness arises partly from observations such as those mentioned above and partly from an improved understanding of the physical basis of the climate system and its changes in the past. In particular, climatologists have become increasingly concerned that human activities could lead to local, regional and, perhaps, global climatic changes, especially in view of the projected increases in population and demand for energy. Our understanding of the climate system is not yet good enough to make detailed predictions of future climatic changes due to natural or anthropogenic causes. However, existing knowledge is enough to be able to suggest the orders of magnitude of potential climatic changes and to point to the parts of the climate system that we should be monitoring carefully.

Probably the largest potential human impact on climate is that due to energy conversion and use. The climate could be influenced, for instance, by releases of waste heat at power plants, by releases of CO_2, SO_2, particles and other climatically important gases from fossil-fuel combustion, or by changes in certain characteristics of the earth's surface, such as its reflectivity, roughness, or wetness.

The possibility of undesirable, and perhaps even irreversible, climate changes due to energy conversion and use could represent a constraint on the future development of energy systems. This potential constraint was examined within the Energy Systems Program of the International Institute for Applied Systems Analysis (IIASA) and the results of this study form the basis of this book. The

1

findings of this Program's study of the global energy system have been published elsewhere (Energy Systems Program, 1981); after developing an understanding of the energy problem, it designed a set of energy models and two scenarios that were used to generate conclusions about energy supply and demand to the year 2030.

This IIASA study relied on population estimates prepared by Keyfitz (1977). There are 4,000 million people on earth today, and he estimated that there will be 4,000 million people by the year 2030. He assumed that the average population replacement would have come down to one by 2015. His population projection indicates that the population growth curve will become flat by 2030.

However, there will be different population growth rates in different parts of the world, with more growth in the poorer countries than the richer ones. Currently the global primary* energy consumption average is about 2 kWyr/yr *per capita*. Consumption is, however, unevenly distributed. Roughly 70% of the world's population consume less than the average and most of this 70% use only 0.2 kWyr/yr of commercial energy (Energy Systems Program, 1981). In 1975, the energy consumption *per capita* in North America was 11.2 kWyr/yr. In Western Europe, Japan, Australia, and New Zealand the average *per capita* energy consumption was 4.0 kWyr/yr. In contrast, the consumption in Latin America, for example, was 1.0 kWyr/yr *per capita*. Clearly, any smoothing of this distribution that leaves the high *per capita* energy consumption untouched would lead to an increase in the average *per capita* consumption.

The IIASA Energy Systems Program considered increases in the average primary energy consumption value from 2 kWyr/yr *per capita* to 3 and 5 kWyr/yr *per capita*. Figure 1.1 shows that, if these estimates are combined with the projected population growth, the total global primary energy demand in 2030 will be 16, 24, and 40 TWyr/yr (1 TW = 10^{12}W = 10^{9}kW), respectively. Thus, the range of 16 to 40 TWyr/yr appears possible in about 50 years from the present, that is, a global primary energy demand between two and five times greater than that in 1975. These estimates can be compared with the estimate made by World Climate Programme (1981) of 27 $\pm$ 25% TWyr/yr.

Table 1.1 shows the global primary energy supply figures for 1975 as derived by IIASA (Energy Systems Program, 1981). Roughly 8.2 TWyr/yr of commercial energy and roughly 0.6 TWyr/yr of non-commercial energy (e.g. fuelwood and agricultural waste) were consumed. The commercial primary energy supply was based almost entirely on fossil fuels.

Table 1.2 shows the two energy supply scenarios developed for the period 1975–2030. In the high scenario, the global primary energy supply in 2030 is 35.65 TWyr/yr, whereas in the low scenario, the supply total in 2030 is 22.39 TWyr/yr. In 1975, 90% of the global primary energy was supplied by fossil fuels (oil, gas, and coal). In the high scenario, the fossil fuels supply 69.5% of the global primary energy in 2030, while light water reactors and fast breeder reactors

---

* Primary energy is the energy recovered from nature: water flowing over a dam, coal freshly mined, oil, natural gas, natural uranium. See the Appendix for a discussion of the energy units used in this book.

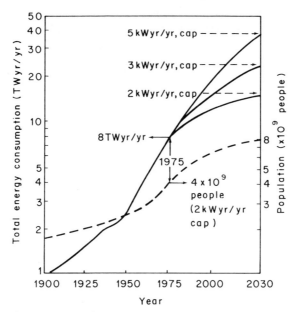

Figure 1.1 Total energy consumption, 1975–2030: three possibilities. The solid lines indicate energy consumption; the dashed line indicates world population.
Source: Energy Systems Program (1981)

Table 1.1 Estimated global primary energy supply, 1975. Source: Energy Systems Program (1981)

Source	Level (TWyr/yr)
Oil	3.62
Gas	1.51
Coal	2.26
Light water reactor	0.12
Fast breeder reactor	0
Hydroelectricity	0.50
Solar[a]	0
Other[b]	0.21
Total	8.22

[a] Includes mostly 'soft' solar individual rooftop collectors; also small amounts of centralized solar electricity.

[b] Includes biogas, geothermal, and commercial wood use and bunkers for international shipments of fuels.

Table 1.2   Two scenarios for global primary energy supply in 2030.
Source: Energy Systems Program (1981)

Source	Level (TWyr/yr)	
	High scenario	Low scenario
Oil	6.83	5.02
Gas	5.97	3.47
Coal	11.98	6.45
Light water reactor	3.21	1.89
Fast breeder reactor	4.88	3.28
Hydroelectricity	1.46	1.46
Solar[a]	0.49	0.30
Other[b]	0.81	0.52
Total	35.65	22.39

[a] Includes mostly 'soft' solar individual rooftop collectors; also small amounts of centralized solar electricity.
[b] Includes biogas, geothermal, and commercial wood use.

together supply 22.7% of the primary energy (in comparison with 1.5% in 1975). In the low scenario, the fossil fuels supply 66.7% of the global primary energy in 2030, and the nuclear sources supply 23.1%.

As the IIASA Energy Systems Program (1981) points out, both the high and low scenarios are basically fossil in nature, which might appear surprising in view of the many statements made at present about our running out of fossil fuels. However, the IIASA Energy Systems Program (1981) suggests that fossil fuels will continue to be available, but will become unconventional and expensive. The transition to a sustainable energy system, that is, one not based on the consumptive use of resources, was found to occur only after the 50-year time frame of the IIASA study. It was not possible to allocate a higher role for nuclear and solar power over the 50-year period; energy production during this period is constrained mainly by the rate of build-up of production capacities, and not by resource scarcity. Nevertheless, the Energy Systems Program (1981) suggests that a transition to a non-fossil world without consumptive use of resources must take place after 2030, when resources become an active constraint.

As for the possible effect of energy systems on climate, several main points can be drawn from the IIASA study. Firstly, the energy demand projections suggest that demand 50 years from now is likely to be two to five times larger than at present. Secondly, during this 50-year period fossil fuels (including unconventional ones) will have to be relied on. Thirdly, since the earth's fossil fuel resources are finite, a transition will have to be made to nuclear and solar power for the longer term.

Thus the question arises: Could the large-scale development of fossil-fuel, nuclear, and solar-energy systems have an effect on the climate? Much of the remainder of this book is devoted to answering this question. It examines the nature and extent of potential climate changes.

In order for the non-climatologist to understand the evaluation of climatic effects of energy systems presented in later chapters, Chapter 2 describes the physical basis of the climate system and methods for assessing the influence of human activities. One of the main assessment methods involves using numerical models of the climate system; generally, they employ computers to simulate the climate of the present day and then to simulate the climate with a perturbation due to energy systems (e.g. waste heat addition, atmospheric CO_2 increase, or widespread reflectivity changes). Estimates of the sensitivities of a model climate to given perturbations can be made in this way, but, as is shown in Chapter 2, model shortcomings mean that it is presently not possible to *predict* future climate changes due to natural or anthropogenic causes. Although there are model shortcomings, models of the climate system still represent the best available tools for assessing the sensitivity of the climate system.

As the Energy Systems Program (1981) concluded, the CO_2 build-up caused by fossil-fuel consumption is probably the most serious climate issue, and Chapter 3 makes a detailed analysis of this issue. We will see that there are many interacting facets. Firstly, it is necessary to know what the natural sources, sinks, and transfers of carbon are. Basically, there are four main reservoirs of carbon: the atmosphere, the land biota, the ocean, and the sediments (including fossil fuels). If CO_2 released by combustion of fossil fuels were to enter the atmosphere very slowly, then there would be no long-term increase of the atmospheric CO_2 content, because the transfer into the other reservoirs and into the ultimate sink of the deep ocean could keep pace with the fossil-fuel CO_2 addition. Since the combustion of fossil fuel has been increasing at an exponential rate during the past 100 years, the addition of fossil-fuel CO_2 to the atmosphere currently exceeds the rate of removal, and an increase in the atmospheric CO_2 concentration is presently being observed. If the use of fossil fuels continues to increase, it is possible that the atmospheric CO_2 concentration might reach a value twice as high as the preindustrial value, or even higher. In recent years this possible CO_2 increase has been much discussed because of the impact of CO_2 on global temperatures and potentially on the global climate system. All other factors remaining constant, a doubling of the atmospheric CO_2 content is presently estimated to lead to an increase in the global average surface temperature of 1.0 to 3.5 K. However, for evaluations of social, agricultural, environmental, and other effects, a knowledge of regional changes of temperature and rainfall is required. Chapter 3 shows that the assessment of regional climatic effects of CO_2 is at a preliminary stage and even the values given for global effects are uncertain.

In view of the potential climatic effects of increases in CO_2 concentration, Chapter 3 also reviews the possible responses to a CO_2-induced climatic change, including the reduction of use of fossil fuels, and compensation for 'damages' and adaptation. In view of the uneven distribution of the effects of increases in CO_2, the CO_2 issue can be seen to be a potential global problem for which new types of solutions must be developed.

The second way in which energy systems can influence the climate, through the release of waste heat, is discussed in Chapter 4. A brief review shows that present-day power plants can have an effect on local conditions, especially with regard to fog formation, cloudiness, and other local humidity characteristics. However, it appears that the climatic effects of the present scale of power plants can be minimized to an acceptable level by correct cooling-tower design and spacing. However, concern arises about the proposals for constructing 'power parks' or conglomerations of power plants producing of the order of 20,000 to 50,000 MW. Studies referred to in Chapter 4 suggest that such power plants could have regional climatic effects, with a potential to enhance or trigger severe weather events, such as whirlwinds, thunderstorms, or tornadoes.

The IIASA studies of the effect of waste heat emphasized the potential impact of large releases of waste heat from point sources. This work involved applying a numerical model of the Northern Hemisphere general atmospheric circulation developed by the United Kingdom Meteorological Office. A number of model simulations were made to examine the sensitivity to various amounts, forms, and locations of waste heat input. It was concluded that waste heat release on the scale foreseen for the next 100 years (i.e., 16 to 40 TWyr/yr primary energy supply in 2030) will not have an effect on the global scale climate, although regional and local scale effects are possible.

Chapter 5 discusses the effects on climate of particles and gases other than CO_2. The influence of particles is extremely complex, since it depends on such factors as the location of the particles both geographically and vertically, and the radiative characteristics of the particles and the underlying earth's surface. Because of this complexity, it is not possible at present to predict the climatic consequences of an increase in anthropogenic particle loading of the atmosphere. In order to make such predictions, detailed models of the atmosphere–ocean system will be required, in which such aspects as the interactions between particles, cloudiness, and radiation are handled in more detail than in present models. Atmospheric circulation must be taken into account, because it seems clear that particles could alter the horizontal and vertical temperature gradients with resulting effects on atmospheric flow.

Other gases are released as a result of energy conversion. Those with a potential influence on climate include sulphur dioxide, oxides of nitrogen, methane, and ammonia. Only preliminary studies with simple models have been made of the potential climatic effects of these gases. It has been noted, however, that many of the so-called trace gases added to the atmosphere by human activities would have a warming effect on the earth's surface similar to that of CO_2.

The effect on climate of a wide variety of renewable energy systems is discussed in Chapter 6. It points out that large-scale deployment of certain technologies could have a climatic effect through changing the earth's surface characteristics over large areas. In particular, it discusses changes in the surface energy balance, roughness, and wetness. A set of model simulations investigated the effect on regional meteorological conditions of a large solar thermal electric conversion plant; they showed that the perturbation due to the solar facility caused changes in the simulated cloudiness and precipitation.

The potential climatic effects of photovoltaic, hydropower, wind power, biomass, solar-satellite power, and ocean-thermal electric conversion systems are also discussed in Chapter 6. It concludes that, owing to the wide variety of renewable energy systems and the different locations that are suited to different technologies (e.g., rainy mountainous areas to hydropower, windy areas to wind power), there is unlikely to be either exclusive development of one technology or a very large-scale concentration of one technology in a given region. It is more likely that a very heterogeneous, distributed use of renewable energy conversion systems will develop. In this case it is unlikely that climate effects would occur on the global—and in most cases even on the regional—scale. Local or microclimatological effects (e.g., over photovoltaic rooftops or immediately downstream of windmills) could occur, but would not be of global concern.

Chapters 3–6 consider the effects of energy systems on climate as a result of additions of CO_2, waste heat, particles, and other gases, and through changes in the earth's surface conditions. In contrast, Chapter 7 considers the influence of climate on energy systems, an interaction that has not received much attention so far in the scientific literature. However, the increasing awareness of the vulnerability of human society to climatic disruption suggests that this interaction will be studied in more detail in the future. As Chapter 7 shows, there are many ways in which climate can influence energy supply and demand. On the supply side, climate data are essential for evaluating renewable energy resources such as solar energy, wind energy, and hydropower. Climate also influences siting of power plants, transporting energy (e.g. shipping or highways), and exploring new energy sources. That the demand for energy is also influenced by climate has been well illustrated by the recent cold winters in the United States. The value of weather and climate forecasts for managing energy systems is also shown in Chapter 7. Finally, the possibility of intentional climate modification for the benefit of energy systems is discussed in Chapter 7, where we see that a much better physical understanding of the climate system and predictive capabilities is required before any kind of intentional climate modification would be feasible.

Chapter 8 summarizes again the main themes of the book and lists the main conclusions drawn in Chapters 2–7. From this summary one can make an evaluation of the present state of the art of assessing the interactions between energy systems and the climate. As for the effects on climate, I conclude that the release of CO_2 by fossil-fuel combustion has the largest potential influence on global climate in the foreseeable future. Nevertheless, there are many uncertainties that must be removed or reduced before detailed predictions can be made of the effects on climate of various energy strategies.

REFERENCES

Energy Systems Program (1981). *Energy in a Finite World: Volume 1: Paths to a Sustainable Future; Volume 2: A Global Systems Analysis.* Ballinger Publishing Company, Cambridge, Massachusetts.

Keyfitz, Nathan (1977). Population of the world and its regions, 1975–2050. WP-77-7, International Institute for Applied Systems Analysis, Laxenburg, Austria.

World Climate Programme (1981). On the assessment of the role of CO_2 on climate variations and their impact. Joint meeting of experts in Villach, Austria, November 1981.

CHAPTER 2

Introduction to the Climate System

2.1 CLIMATE FROM THE POINT OF VIEW OF ENERGY SYSTEMS' ANALYSIS

In order to be able to assess the potential effect on climate of energy conversion and use, it is necessary to have a basic physical understanding of what climate is, how it changes, and the tools available for studying it. In one chapter such information can only be conveyed briefly and incompletely. However, this chapter considers some of the main features of the climate system that are most relevant for evaluating the influence on climate of energy systems.

Firstly, the climate system as a whole is described. Although it is common to think of the 'climate of an area' as the long-term average temperature and rainfall, climatologists consider the climate system to be the combination of all the variables within the earth–atmosphere envelope; ice, water, air, land, living matter, and all their characteristics. Because of the many interacting components, the natural climate system is very complex and presently not completely understood.

Secondly, some observations of the natural variations of climate are described, in order that the potential changes due to energy systems can be kept in perspective.

Considerable attention is also given in this chapter to the tools available for studying the sensitivity of the climate system. Particular emphasis is placed on the use of numerical models of climate, since they have been used widely to increase our understanding of climate and are a potential method for climate prediction.

2.2 THE NATURE OF THE CLIMATE SYSTEM

The climate system as a whole consists of five interacting subsystems: atmosphere, ocean, cryosphere (ice and snow), biomass, and land. As Figure 2.1 shows, numerous processes, such as evaporation and heat exchange, link these components. Because of all the interactions, the total climate system is very complex and changes in one part of the system can lead to changes in another part.

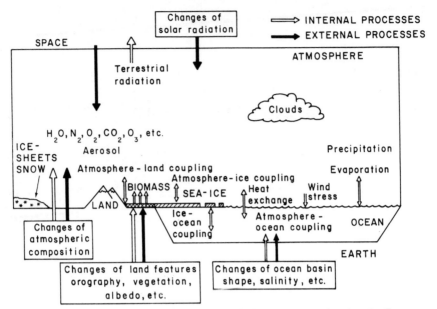

Figure 2.1   A schematic representation of the climate system, showing the five components (atmosphere, ocean, land, ice, and biota) and some of the processes that link them. Source: GARP (1975)

As a convenient approach to considering climate variation, the climate system can be divided into an internal and an external system. The internal system consists of the air, water and ice envelopes around the earth and the external system consists of the underlying ground and the outer space.

Climatic variations can be caused, on the one hand, by changes of the external system; for example, by changes in the earth's crust, such as continental drift, or by changes in the orbit of the earth around the sun or the amount of solar radiation arriving at the outside of the atmosphere.

On the other hand, climatic variations can be caused by changes in the internal system, which are manifested as feedbacks between the variables within the air, water, and ice layers. For example, there is a feedback between the atmosphere and ice, as illustrated in Figure 2.2. In this case, an increase in temperature leads to a decrease in ice/snow extent with a consequent decrease in albedo (reflectivity). This decrease means that more solar radiation can be absorbed and a further warming takes place, giving a positive feedback loop. The relation between surface temperature and cloudiness provides an example of a negative feedback loop. If one starts with an increase of surface temperature due to an increased absorption of solar energy at the surface, then this increase can lead to increased evaporation from the surface and thus to enhanced cloud building. The cloudiness reduces the amount of solar radiation reaching the earth's surface and thereby reduces the surface temperature. Although this is an example of a negative feedback loop, involving the variables of temperature, cloudiness, and incom-

ALBEDO-POLAR ICE COVER-TEMPERATURE
FEEDBACK LOOP

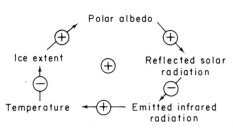

Figure 2.2   The feedback loop between surface
temperature and ice or snow cover. Source:
Kellogg (1975)

ing solar radiation, in reality many other variables, such as long-wave radiation or cloud height, are involved, and the complete nature of the cloudiness–temperature feedback loop is not known.

Many of the feedback loops within the climate system have been identified (e.g., Schneider and Dickinson, 1974) and the concept of a feedback process is an extremely important one in the study of climatic variations. Unfortunately, the interactions among the many feedback processes acting simultaneously are not completely known. While an understanding of most of the climatic feedback loops and their interactions is necessary for the analysis of the potential effects of man's activities on climate, the observational and model sensitivity studies described in this chapter do contribute to an improved understanding of the physical basis of climate, that is, of how the components of the climate system interact.

2.3   A DEFINITION OF CLIMATIC STATE

The boundaries between what is called 'weather' and what is called 'climate' are difficult to draw. Section 2.2 described the climatic *system* briefly, and we saw that climatic *variables* are, for instance, atmospheric temperature, wind speed and direction, ocean surface temperature, snow depth, and soil moisture amount. For the purposes of studying natural and man-made climatic variations, using models or observations, it is useful to have a definition of climatic state. The United States National Committee for GARP (1975) defined climatic state as the average (together with the variability and other statistics) of the complete set of climatic variables (i.e., atmospheric, hydrospheric, and cryospheric variables) over a time interval and over a domain of the earth–atmosphere system, which must both be specified. The time interval refers to a period *longer than* the life span of individual weather systems (of the order of several days) and longer than the time limit over which the behaviour of the atmosphere can be predicted locally (of the order of one to three weeks). In practice this means that the climatic state is defined for periods of a month or more and thus the January, winter, annual, decadal, etc., climatic states can be considered.

2.4   NATURAL VARIATIONS OF THE CLIMATIC STATE

ct and indirect observations of climatological variables, such as temperature, precipitation, and pressure, show that climatic state, as manifested by these variables, varies on all time scales. Observations have been made with instruments (thermometer, barometer, etc.) since the 17th century, although only in the last 30 years has the network of meteorological stations given enough coverage that estimates of changes in hemispheric seasonal average surface temperature can be made with some confidence. Observations of variables in the atmosphere above the earth's surface cover a limited period of 20 to 30 years, with a limited number of meteorological stations mainly in the Northern Hemisphere.

In order to study the climatic state during periods before the time in which instrumental observations have been recorded, proxy records must be used. These include tree rings, pollen in peat deposits, ocean bed cores and ice cores, all of which can give climatic information. Fritts (1976) has described using tree rings to derive climatic information. It is also possible to reconstruct climatic history from documented historical data, such as dates of grape harvests and floods, prices of grains, freeze and thaw dates for lakes and rivers, etc. (see, for example, Lamb, 1977; Wigley *et al.*, 1981). Figure 2.3 shows schematically how climate is understood to have varied during the last 20,000 years, as derived from a variety of evidence.

During the last 1 million years a series of glacial and interglacial periods occurred. The earth is at present in an interglacial period, which began about 10,000 years ago. At the maximum of the last glacial period, about 20,000 years ago, ice sheets covered large areas of North America and Europe, for instance, with global temperature several degrees lower than that of today. After the peak

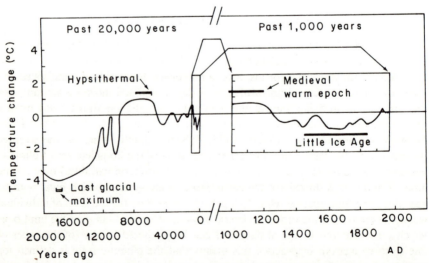

Figure 2.3   Schematic representation of temperature changes during the past 20,000 years. Source: Wigley (private communication, 1982)

of the last glacial period, the continental ice sheets receded and
recorded. A period, referred to as the 'Altithermal', or 'Hypsitherr
which temperatures were warmer than today. This postglacial v
its peak at different times in different places between 4,000 and 8,ᴜᴜ᷉
Since this period, a cooler period is recorded about 2,300–2,900 years ago a᷉
then the mild conditions of the maximum postglacial warmth were nearly reached
once more within the medieval warm period (800 to 1200 AD). During this warm
time, ice conditions around Iceland and Greenland were much less severe than
today, in Western and Central Europe the cultivation of the grape vine extended
3–5° latitude further north and it is estimated that the average summer
temperature was about 1 °C higher than today. Between the 16th century and the
middle of the 19th century, a comparatively cool period, commonly referred to as
the 'Little Ice Age', occurred in which mountain glaciers and snow cover were
expanded compared with the present situation.

The above brief, simple description of natural climatic fluctuations in the past
can be used to put into perspective the potential climatic fluctuations caused by
man's activities, including energy conversion. Because of the complexity of the
climate system and also because of inadequate observations in terms of spatial
and temporal coverage, the causes of the natural climatic variations have not been
determined to the extent that predictions of future variations can be made. The
knowledge of past climate is based on information on temperature and precipitation (and, rarely, other variables such as wind direction) from a limited number of
sites.

2.5   CLIMATIC VARIATIONS DURING THE LAST 100 YEARS

During the past 100 years, meteorological observations have been made at an
increasing number of stations around the globe. The nature of climatic fluctuations during this period has recently been studied in detail. The accumulated information is useful for an improved understanding of climatic changes and thus for
estimating the sensitivity to man-made perturbations.

Figure 2.4 shows filtered seasonal and annual mean surface temperatures for
the Northern Hemisphere (Jones *et al.*, 1982). The curves show an early 20th
century hemispheric warming that has been noted by many authors. This
Northern Hemisphere warming peaked in the late 1930s in annual, spring,
summer, and autumn values. In winter the warming continued until the mid-
1940s. Then a general cooling set in that continued until the 1960s in annual
values and in all seasons and to the mid-1970s in autumn temperatures. In recent
years there has been a renewed warming trend, but the start of this warming
varies considerably from season to season.

Considering the period 1942 to 1972, which is one of general cooling at the
surface in the Northern Hemisphere, a first point to note is that different methods
of compiling the observations and deriving trends produce different numerical
values for the average temperature change during the period (van Loon and
Williams, 1976). Secondly, although the period as a whole was one of cooling,

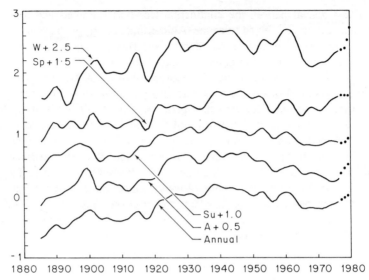

Figure 2.4   Filtered seasonal and annual mean surface temperatures for the Northern Hemisphere. The ordinate (°C) and the curves have been displayed for clarity. The dots show estimates for recent years. Source: Jones *et al.* (1982)

warming did occur during some of the time. van Loon and Williams (1976) have shown for the region 15–80°N that in the winter season the temperature change during the period 1942–1972 was –0.21 °C, while during the period 1950–1964 it was +0.39 °C. Thirdly, each of the seasons experienced slightly different magnitudes of temperature change, as shown in Table 2.1. Fourthly, it has been found, not surprisingly, that regional variations are different from the hemispheric or global mean value. The change during 1942–1972 of zonally averaged (i.e., averaged for all points around a latitude circle) temperature in each season is illustrated in Figure 2.5. During each season the changes in low latitudes (15–30°N) were not large. In winter, spring, and autumn the largest changes were in the polar latitudes, while in summer the largest changes were in middle latitudes. The difference in regional distribution of temperature change according to season suggests that different processes are responsible for the changes in the

Table 2.1   Average temperature change in each season during the period 1942–1972 averaged over the area 15–80°N. Source: Williams and van Loon (1976)

Winter	–0.21 °C
Summer	–0.19 °C
Spring	–0.36 °C
Autumn	–0.29 °C
Year	–0.26 °C

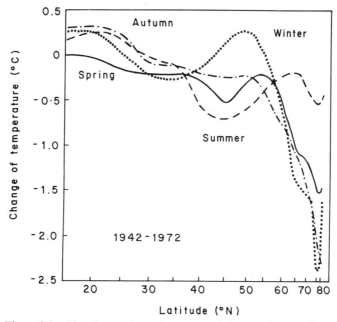

Figure 2.5   The change in each season of the zonal mean (i.e., latitude average) temperature (°C) between 1942 and 1972. Source: Williams and van Loon (1976)

different parts of the year and, consequently, consideration of annual averages will be misleading.

Figure 2.6 shows the isopleths of the slope of the linear trend of winter mean temperature during the period 1942–1972. This map illustrates the fact that a small hemispheric change can result from the occurrence of large and opposite regional trends. The hemispheric cooling is largely a result of the occurrence of two extensive areas of strong cooling in polar latitudes. In the period 1900–1941, which was one of hemispheric warming, the largest trends in the winter season were again noted in the polar area of the Northern Hemisphere.

Since the size and sign of temperature changes over the oceans in the tropics and in the Southern Hemisphere is undetermined because of the poor observational network in these areas, the amplitude of *global* trends is certainly open to question, but even their sign may be considered uncertain (van Loon and Williams, 1976). It should also be pointed out that the above analyses refer only to surface temperature and not to temperature at other levels of the atmosphere. An analysis of temperature changes during the period 1949–1972 at the 700 mbar level in the winter season (van Loon and Williams, 1977) has shown that the zonally averaged changes had quite a different distribution from those at the surface during the same period, with significant implications regarding associated changes in the general atmospheric circulation.

Surface temperature is only one of many variables needed to describe the

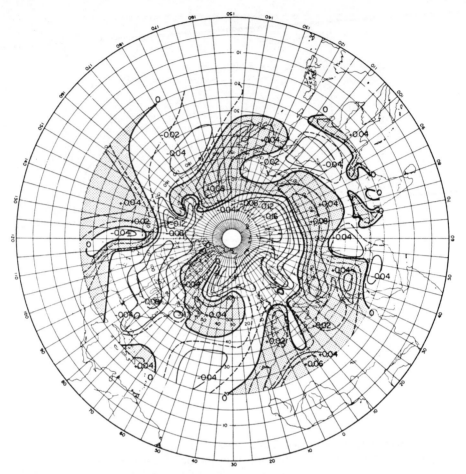

Figure 2.6   The isopleths of the slope (°C/yr) of the linear trend line computed for the period 1942–1972 for available meteorological stations in the Northern Hemisphere. Shaded areas show where the winter temperature increased during this period. Source: van Loon and Williams (1976)

climatic state and its variations. Knowledge, for instance, of the pressure and precipitation and of variables at other levels of the atmosphere and information on the other subsystems (ocean temperature distributions, ice/snow depth and extent) are also required for an understanding of how and why variations of the climatic state occur and for acquiring the ability to predict future natural and man-made variations.

Nevertheless, the available observations are sufficient to demonstrate the nature of climatic variations. Climate is seen to vary on a wide range of time scales, from the interannual to the geological. Within the last 100 years, observations show that global or hemispheric surface temperature changes are a resultant of large, coherent areas of regional change. In winter, spring, and autumn the changes have been larger in polar areas than in middle and low latitudes, while in the summer changes are larger in middle latitudes. Because the network of meteorological stations is poor in some areas, the global trend in surface temperature in recent years is uncertain (Kukla *et al.*, 1977). The causes of the observed changes are not known, although many plausible theories have been put forward (see, for example, Lamb, 1972).

2.6 THE WAYS IN WHICH ENERGY SYSTEMS CAN INFLUENCE THE CLIMATE SYSTEM

(a) Scale

The climate system can be influenced on a range of space scales. On a local scale (<25 km; Orlanski, 1975) energy systems can already be seen to influence climate; differences between urban and rural climates have been well documented and are discussed in Chapter 4.

On a regional scale (25–2,500 km), man's activities are observed to be influencing climate and the potential influence of energy systems is great. For example, the general view now appears to be that precipitation enhancement occurs downwind of urban areas; modifications of temperature, humidity, cloud, and precipitation may persist to as far as 100 km downwind of urban areas (Oke, 1980).

On a global scale there is no indisputable evidence that man's activities are influencing climate at the present time, but such an influence is possible in two ways. Firstly, the global climate could be affected by a change in the concentration of an atmospheric constituent (e.g., CO_2 or water vapour). Such a change, because it would be spread by atmospheric circulation, would occur globally. This does not mean that the resulting climatic variation would be the same over all regions of the globe; a doubling of the atmospheric concentration of CO_2 would not lead to the same surface temperature change in all regions. Secondly, global climatic variations could be caused by a large change of climatic boundary conditions in one area. A large enough anomaly could give rise to a change in the atmospheric (or oceanic) circulation, such that other changes occur upstream and downstream of the area. This is discussed in more detail in Chapter 4. Sawyer (1965) suggested that an anomaly in heating of an area greater than 10^5–10^6 km^2 and magnitude 20 W m^{-2} could influence the atmospheric circulation. Observations and some model studies of the climate system have indicated, for example, that widespread anomalies of ocean surface temperature can cause climate variations elsewhere (see, e.g., Namias, 1975; Bjerknes, 1969; Rowntree, 1972, 1976; and later in this chapter).

(b) Points of influence

There are three main ways in which energy conversion systems can influence the climate system. Firstly, energy conversion, either at a power station or at most points of end use, releases heat into the climate system.

The second way that energy conversion systems can influence the climate system is through changing the concentration of atmospheric constituents. Certain of these play an important role in the absorption or emission of radiation or in chemical reactions, which determine the amounts of other constituents. The addition of CO_2, water vapour and particulates from energy conversion systems (particularly from combustion of fossil fuel) could have significant influences on the radiative properties of the atmosphere and thus on the earth's surface temperature and/or thermal gradients. Other constituents, e.g., sulphur and nitrogen compounds, are also produced by energy conversion (Bolin and Charlson, 1976; Hahn, 1979). The effect of SO_2 from combustion of fossil fuels has already been noted on a regional climatic scale (Weiss *et al.*, 1977).

The third effect of energy conversion systems is through changes in the characteristics of the land (or ocean) surface. Large-scale development of solar thermal electric conversion systems would lead, for example, to widespread changes in the surface roughness (because of the large arrays of heliostats) and, possibly, surface hydrological characteristics (due to paving previously vegetated areas, for example), in addition to changes in the surface heat balance. Little information is available to assess the scale of the resulting climatic changes. Observational and model studies (e.g., Namias, 1960; Walker and Rowntree, 1977) have, however, indicated that widespread anomalously wet or dry surface areas can influence the atmospheric circulation. Ocean thermal energy conversion systems, if used on a large scale, could cause ocean surface temperature and reflectivity changes and, possibly, ocean circulation changes, with possible climatic implications.

The following chapters will discuss these effects of energy conversion systems in more detail. The climatic constraints of the three energy options will also be examined.

2.7   TOOLS FOR STUDYING THE EFFECTS OF ENERGY SYSTEMS ON CLIMATE

For the purpose of evaluating the climatic constraints on the transition from today's energy supply system to that proposed for the future, some kind of prediction is required of the likely effect of the future systems. Because of the nature of the climate, it is not appropriate to experiment in order to assess the effects. For example, in order to evaluate the climate constraints of the fossil-fuel option, there are obvious reasons why mankind could not proceed to double the atmospheric CO_2 concentration with the intention that, if the climatic consequences are undesirable, an immediate return to present conditions could be made. The effects could be irreversible. In order to assess the effects of energy options on climate, before intentionally or unintentionally performing the experiment, two approaches are available. The first is to study situations analogous to those arising from energy systems as a guide and the second is to use models of the climate system.

The use of analogues is useful, for example, in assessing the effect of large releases of waste heat at energy parks (see Chapter 4). Since no energy parks producing 10,000–50,000 MW electricity exist at present, their potential influence can be assessed on the basis of the observed effects of comparable sources of heat and moisture, such as islands heated by solar radiation, urban-industrial complexes, forest fires, and phenomena such as volcanoes (see, for example, Koenig and Bhumralkar, 1974; Rotty, 1974; Hanna and Gifford, 1975). Similarly, Bhumralkar (1977) has suggested that the effects of urban areas on climate can be taken as an analogue for the effects of solar thermal electric conversion plants, since they should have similar thermal and mechanical effects.

The second approach, that of using climate models, has been used increasingly in recent years. The use of equations to describe the climate system is the basis of climate modelling. Because knowledge of the physical system is incomplete and because of computational limitations, both physical and numerical approximations must be made in using equations, and a hierarchy of climate models has resulted. At the lower end of the hierarchy are models consisting of one equation, which describes the climate system in terms of the global balance between the amounts of incoming solar radiation and outgoing long-wave radiation. At the other end of the hierarchy are the complex general circulation models, which simulate the three-dimensional circulation of the atmosphere and ocean, using time-dependent equations. More detailed descriptions of the physical basis of climate and climate models have been given by GARP (1975), Schneider and Dickinson (1974), and Gates (1979), and the use of models in the study of the potential effects of energy systems has been described by Williams (1979). It is not necessary to describe the entire hierarchy or methodology here, and reference is made to other publications. However, three types of models, which have been used in studies of the effect of energy conversion systems on climate, will be discussed below.

(a) One-dimensional radiative–convective models

These models consider the vertical structure of the atmosphere either for globally averaged conditions or for a particular latitude average or point. The model determines the thermal structure of the atmosphere by considering the balance between the radiative flux and a parameterized (i.e., approximation to) convective flux. Therefore, incoming solar radiation, planetary albedo, surface temperature, the vertical distribution of temperature and of optically active atmospheric constituents are considered for the global average or for an average (e.g., middle latitude, polar latitude) location on the earth's surface. The model computes the temperature profile corresponding to the input conditions, adjusting the profile so that the vertical temperature gradient does not exceed the pre-assigned stable value of 6.5 K/km for convection (this is the 'convective adjustment'). Despite the fact that such models ignore the dynamics of the climate system, they have been very useful in showing the order of magnitude of the change in equilibrium surface temperature and temperature profile for different changes in atmospheric constituents (e.g., Manabe and Wetherald, 1967; Rasool and Schneider, 1971; Reck, 1974; Augustsson and Ramanathan, 1977).

(b) Energy-balance models

Energy-balance models are also from the lower end of the hierarchy. They have been used, for example, to assess the effect on globally-averaged temperature of increases and decreases in the amount of incoming solar radiation. Development of these models was pioneered by Budyko (1969) and Sellers (1969). Empirical relations are the basis for this type of climate model. Basically the model considers the heat balance of the earth–atmosphere system:

$$Q(1 - \alpha_p) - I = C + B$$

where

Q is the solar radiation reaching the upper boundary of the earth's atmosphere;
α_p is the planetary albedo;
I is the outgoing long-wave radiation;
C is the heat redistribution due to horizontal movement in the atmosphere and ocean; and
B is the gain or loss of heat in the system.

For mean annual conditions B is zero. It is found from observed data that C is related to $(T - T_p)$, where T_p is the global-average air temperature near the earth's surface and T is the mean air temperature near the surface for different latitude zones. From this relation, together with the equation of the heat balance of the earth–atmosphere system, it is possible to calculate the mean annual temperature at different latitudes. The distribution of this temperature agrees surprisingly well with that observed. It is then possible to investigate how changes in the heat balance (for example, due to waste heat input, or addition of CO_2 to the atmosphere) influence the temperature distribution. Budyko (1971) used the energy balance model to show that a few tenths of a per cent increase in solar energy input or a few tenths of a degree centigrade rise in the mean air temperature at the earth's surface could lead to a complete melting of the polar pack ice in the Northern Hemisphere. However, this result is based on a model that is a very simplified representation of the climate system.

(c) General circulation models

At the top of the climate model hierarchy are the so-called 'general circulation models' or GCMs. For thorough studies of the climate system, GCMs that treat each of the climatic subsystems (i.e., atmosphere, oceans, and ice) are required. At the present time atmospheric GCMs have been developed the most. Atmospheric GCMs predict changes of such variables as wind velocity, temperature, moisture, and surface pressure by solving equations of motion, the thermodynamic equation, and conservation equations for moisture and mass. Most of the GCMs from which results will be described in this book solve the equations for a set of grid points, which have a few degrees latitudinal and longitudinal resolution and a few kilometres vertical resolution. In other models the horizontal distributions of the variables are represented by a limited number of spherical harmonics and the vertical distributions are specified at finite difference levels.

It is found that, when the equations are solved with boundary conditions representing, for example, January of the present day, the models reproduce the basic features of the atmospheric circulation in January quite realistically (temperature, pressure, and wind distributions, for example). The boundary conditions, which must be specified, include the incoming solar radiation, the extent and height of the land surface, the albedos of land surfaces, and the distribution of ocean surface temperatures.

It is recognized, however, that atmospheric GCMs have shortcomings. In particular, the absence of a joint atmospheric–ocean system, poor treatment of clouds and hydrological processes and of subgrid scale processes have been noted. Despite the shortcomings, it is felt that atmospheric GCMs represent the best tools available at the present time for studying the sensitivity of the atmosphere to changes caused by energy conversion systems. Several models, which differ in details of their construction, have been used for such studies.

The way in which model sensitivity studies are carried out varies somewhat from model to model. In a number of model studies described elsewhere (particularly in Chapter 4) the procedure follows a pattern: the GCM is integrated with boundary conditions representing those of the present day (January or July); the atmosphere at the beginning of the model run can be set at rest with a uniform temperature distribution or can be set to represent an observed distribution; the integration, with a time step of the order of 10 minutes, is performed for a simulated time of 60 to 120 days (or, sometimes, longer). After 20 to 30 days the simulated circulation comes into a quasi-steady state. Averages of meteorological variables for the last 30 (or 40) days of the simulation are taken to represent the climate of the control or standard case. The GCM is then integrated with an alteration of boundary conditions to represent the change in external forcing —waste heat is included or the atmospheric CO_2 content is changed. Differences in 30 (or 40) day mean meteorological fields between the control case and the perturbed case of the GCM are then determined.

An important part of this kind of sensitivity study is, however, to determine how much of the difference between control and perturbed cases is due to the inherent model variability. That is, just as the real atmosphere has an inherent variability manifested by the difference between the January climatic state of two years, for example (see Leith, 1973; Madden, 1976), the GCMs also have an inherent variability, which is manifested by the fact that, when a series of control cases are integrated, which differ only by small random differences in the initial conditions, different results are obtained for each case. It is necessary to determine the inherent model variability from a series of control runs (i.e., establish the noise level of the model) so that the significance of the response of the simulated atmospheric circulation to the imposed perturbation (i.e., signal) can be assessed (for detailed discussion of this topic, see Chervin and Schneider, 1976a and b; Chervin et al., 1976; Laurmann and Gates, 1977). Specific examples of the use of GCMs to investigate the response of the simulated atmospheric circulation to additions of waste heat and atmospheric CO_2 and to large-scale changes in boundary conditions will be discussed in subsequent chapters.

As mentioned above, most emphasis has been on the development of atmospheric circulation models. However, in recent years, as an awareness of the important role of the oceans within the climate system has grown, more attention has been given to coupled models of the atmospheric and oceanic circulations. Much work, of course, remains to be done, especially in the development of the ocean models. One sensitivity study with a model considering both the atmosphere and the ocean has been reported (Manabe and Stouffer, 1979, 1980). In this case, the ocean was modelled as a static isothermal layer of water of uniform thickness with provision for a sea-ice layer. The ocean was therefore treated very simply, with such features as ocean currents not considered at the present stage. The ocean layer was assumed to be 68 m thick and thus heat storage associated with the annual cycle of observed sea surface temperature was simulated fairly realistically. Ocean temperature change was calculated by balancing the surface heat fluxes. In addition further processes were modelled in connection with sea ice.

The procedure for making model sensitivity tests is somewhat different with this model from that described above. Rather than simulating one month, the entire seasonal cycle is simulated. The model starts from an isothermal, dry, and motionless atmosphere with an isothermal ocean. Both the control case and the perturbed case are integrated over several years of model time. In tests described in more detail in Chapter 3, the model produced stable climatic conditions in about a decade of model time and the control case was found to simulate the observed basic characteristics of both geographical and seasonal temperature variations successfully. The control and perturbed cases are compared by computing in each case a mean annual average over the last three-year period of the simulation.

Schlesinger and Gates (1981) have examined the role of the ocean in the global climate system in a series of five simulations with a two-level atmospheric GCM. In the control experiment, the sea surface temperature and sea-ice distributions were given observed long-term monthly-averaged values. The other four experiments considered annually averaged sea surface temperature, halved oceanic evaporation, a slab ocean mixed layer, and no ocean whatsoever. In the last case it was assumed that most of the grid points that are usually treated as ocean had the most ubiquitous low-elevation continental surface type for each latitude.

In the 'no ocean' experiment the surface air temperature was much higher than at present in the summer hemisphere and much colder in the winter hemisphere. Since the atmosphere's moisture source was effectively removed in this case, the precipitation and cloudiness were greatly reduced. When the annual mean sea surface temperatures were assumed, the surface air temperature and precipitation distributions became more summer-like in the winter hemisphere and more winter-like in the summer hemisphere. The halved oceanic evaporation reduced the atmospheric moisture content and relative humidity, but there were no major changes in the global patterns of precipitation and temperature.

In the simulation where the ocean and atmosphere could interact, it was assumed that the mixed layer was a slab of 60 m thickness with no fluxes through its base. The results of this simulation showed large-scale differences of sea surface temperature in comparison with the observed distribution used in the control experiment, especially in the equatorial and eastern oceans. The surface air temperature and precipitation were also different from those in the control case. Schlesinger and Gates (1981) concluded that a proper simulation of the sea surface temperature is necessary for the successful simulation of climate. The results of the 'slab mixed layer' experiment further indicated that a successful simulation of the sea surface temperature will at least require incorporating the dynamics of the surface ocean.

(d) Past climatic states

A further tool for studying the effect of energy systems on climate is the record of past climate. It has been determined, for example, from state-of-the-art climate models that a doubling of the atmospheric CO_2 concentration would lead to an increase of the earth's surface temperature of 2–3 °C. However, the models still have shortcomings that make it impossible to describe in detail the climatic consequences of the CO_2 increase. Consequently, Kellogg (1978) and Flohn (1980) have suggested that information could be derived from knowledge of past epochs when the average temperature was warmer than that at present. Both Kellogg and Flohn have considered the evidence from the period of maximum postglacial warmth as a guide to regional changes in precipitation that might occur with temperatures 2–3 °C higher than current temperatures (see Chapter 3). This approach is difficult, however, because of the time-transgressive nature of such epochs. Since the peak warmth did not occur everywhere at the same time, the regional precipitation anomalies associated with the maximum postglacial warmth also were not simultaneous. It is also difficult because, as Flohn points out, man has recently changed climatic boundary conditions (vegetation cover, albedo, soil moisture, composition of the air, etc.); thus, it is reasonable to doubt that climatic history can repeat itself. Lastly, the cause of the maximum postglacial warmth probably differed from that of the potential future warming and there is assurance that the highly non-linear climate system would respond in the same way to both impulses, even if they are similar.

Two recent studies have used instrumental observations of temperature, rainfall, and pressure during the present century as a basis for discussing the response of the climate system to a warming (Williams, 1980; Wigley *et al.*, 1980). Williams (1980) has looked at regional rainfall, temperature, and pressure anomalies in the Northern Hemisphere for seasons within the last 70 years when the Arctic was warm. The reason for choosing warm Arctic seasons is a result of the model and observational studies, which have suggested that the Arctic is more sensitive to climatic changes. Clearly, the warm Arctic seasons were not a result of a CO_2 increase or other anthropogenic factors, they were rather the result of non-linear interactions between components of the climate system. The relevance of looking at warm Arctic seasons is that they show that:

● large and coherent anomalies of temperature occur elsewhere in the hemisphere when the Arctic is warm, but these anomalies are not all positive;
● related changes in the atmospheric circulation lead to large, coherent anomalies in the rainfall distribution;
● the magnitudes and distribution of the anomalies are different in the different seasons.

A related approach, using past instrumental records of climate to gain insight into possible futures, has been developed by Sergin (1980). In this 'similarity method' seasonal variations in the fields of meteorological variables are taken as a physical model of climatic variations in the average annual field. Using this method, Sergin has investigated the results of a warming and a cooling of the average annual surface temperature by 2 °C. In describing this work, Ausubel and Biswas (1980) have also pointed out that, in making use of natural fluctuations, we are assuming that, given similar boundary conditions, there are broad similarities in the way the atmosphere responds to different types of forcing; this is the basis of the scenario approach. The latter has been found to be true for certain model experiments. Manabe and Wetherald (1980) find that the response of their GCM to a doubling (or quadrupling) of the CO_2 content is very similar to the corresponding response to a 2% (or 4%) increase in the solar constant (i.e., the amount of solar radiation reaching the top of the atmosphere).

(e) Concluding remarks

For each of the approaches to evaluating the effect on climate of energy conversion systems, analogues, climate models and climate history, there are limitations. The evaluation cannot be made by one method alone; a combination will have to be used, depending on the problem being investigated and the available information. In any event, a prediction of the effects on climate cannot be made with the presently available tools, but information can be gathered for a general evaluation of the climatic constraints on different energy systems.

2.8 OCEAN SURFACE TEMPERATURE ANOMALIES AND CLIMATE

This section will review briefly some of the observational and model studies of the effect of ocean surface temperature anomalies on climate. The key reason for presenting such a review is that it illustrates the nature of the interactions between two major components of the climate system.

When the ocean surface temperature over a large area differs by several degrees from its long-term average temperature, the overlying atmosphere and sometimes the atmospheric circulation over a wider area can be affected. The effect of ocean temperature anomalies on climate could be taken as an analogue for the effect of certain energy systems on climate, for example, power plants situated in ocean areas could warm the surface water and influence the atmospheric circulation, or ocean thermal electric conversion plants (see Chapter 6) could cool the surface water and influence climate.

As Barnett (1981) has pointed out, there are two conflicting views of the role of the oceans in the climate system. One line of reasoning holds that the ocean provides a stabilizing influence in what would otherwise be a more chaotic climate system. The other point of view is that, because the ocean is a heat and moisture source for the atmosphere, it can influence effectively future climatic variability.

Barnett (1981) has reviewed studies of the role of the Pacific Ocean in the climate system. Namias (1975) and others have demonstrated a close simultaneous relation between sea surface temperature anomalies in the mid-latitude north Pacific Ocean and overlying atmospheric patterns. Relations between anomalies of the atmospheric pressure field over the Pacific and 'downwind' effects over North America were also established clearly.

Efforts to verify these empirical studies using simple and complex general circulation models have, according to Barnett (1981), not given clear results. Some of the model studies are described below.

The tropical regions of the Pacific Ocean are also thought to have an influence on the atmosphere. Bjerknes (1966, 1969) described interactions between the sea surface temperatures in the tropical Pacific Ocean, rainfall, and subsequent release of latent heat that influences the atmospheric circulation. Empirical evidence supports the hypothesis that the tropical Pacific Ocean can significantly influence mid-latitude (North American) climate. Studies made with general circulation models (GCMs) have suggested also that there is a strong response in the atmosphere to changes in tropical oceanic boundary conditions. The studies of Rowntree (1972) and Julian and Chervin (1978) are briefly described below.

Barnett (1981) lists the few attempts that have been made to predict subsequent atmospheric behaviour using ocean variables. He then proceeds to investigate the hypothesis that changes in sea surface temperature in the Pacific Ocean can be used to predict subsequent changes in climate over the North American continent. In particular, he attempted to determine the relative predictive role of changes in the sea surface temperature (SST) of the mid-latitude and equatorial Pacific, respectively. As shown in Figure 2.7, one set of predictors used in Barnett's study was areally averaged sea surface temperatures for key areas. Barnett concluded that prior knowledge of SST anomalies in the Pacific can be used to hindcast air temperature anomalies over North America. The best success occurred for winter and the poorest for summer. The skill levels were, however, uniformly low in all seasons. Over the central part of North America significant skill could not be found for any season. Approximately half of all predictive skill came from tropical SSTs, with the most important predictor region (SST 5; see Figure 2.7) being off the coast of South America. The second most important predictor region (SST 6) was along the coast of North America. Barnett also found that the nature of the equatorial SST patterns in spring through autumn was particularly important in

26

SST predictor regions

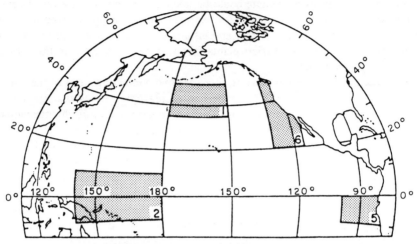

Figure 2.7   Location of sea surface temperature (SST) predictor regions in the Pacific Ocean. Source: Barnett (1981)

estimating the development of winter conditions over North America, but not summer conditions.

Figure 2.8 shows the ability of SST anomalies to hindcast fluctuation in surface air temperature in winter one season in advance. The contours show the per cent of variance accounted for by the linear prediction model developed by Barnett using the SST data. The predictability is highest along the west coast, the southeastern United States, the northern states and southern Canada. Barnett points out that the prediction models based on SSTs seldom explain more than 20% of the observed variance in air temperature anomalies.

A considerable number of experiments have been made with atmospheric GCMs to investigate the response of the simulated atmospheric circulation to anomalies in the ocean surface temperature. The results of these experiments have contributed significantly to the understanding and modelling of the climate system. Some studies have, as mentioned above, considered the effect of anomalies in the eastern equatorial Pacific Ocean (e.g., Rowntree, 1972; Julian and Chervin, 1978) and have shown that when the model boundary conditions include warm temperatures in this area the changes in the model atmosphere in many cases resemble those observed in the atmosphere.

Rowntree (1972) tested the sensitivity of a Northern Hemisphere GCM to ocean surface temperature increases in the eastern equatorial Pacific and found a considerable model response, including rising air over the temperature increase, precipitation increases west of the anomaly, and a strenghtening of the subtropical jet stream in the eastern Pacific. Julian and Chervin (1978) used a global GCM to look at the response to ocean surface temperature anomalies in the tropical Pacific Ocean and also found that many of the observed atmospheric variations

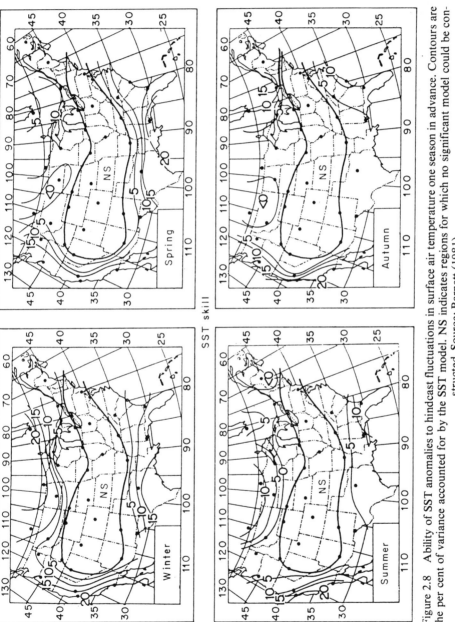

Figure 2.8 Ability of SST anomalies to hindcast fluctuations in surface air temperature one season in advance. Contours are the per cent of variance accounted for by the SST model. NS indicates regions for which no significant model could be constructed. Source: Barnett (1981)

were reproduced. An important feature of both model and observational studies is that temperature variations in the eastern equatorial Pacific Ocean can have such far-reaching effects, because the large-scale atmospheric circulation, which transports heat from the equatorial to the polar areas, is influenced.

Several tests of the sensitivity of the GCM of the National Center for Atmospheric Research (NCAR) to prescribed changes in ocean surface temperature have been made. Kutzbach *et al.* (1977) investigated a configuration of ocean surface temperature anomalies in the north Pacific Ocean, as illustrated in Figure 2.9. In one experiment the anomalies were three times larger than those illustrated. Figure 2.10 shows the geographical distribution of differences in the pressure field at 6 km between a simulation of normal January conditions and a simulation with the ocean temperature anomaly. It can be seen that the pressure did not change only in the immediate vicinity of the temperature change. Kutzbach *et al.* (1977) concluded that the GCM exhibited a well defined response to large changes in mid-latitude ocean surface temperature. Cyclones formed and/or intensified over warm ocean surface temperature anomalies and tended to be weak or absent over cold anomalies. Many features of the response to the large prescribed changes were statistically significant. Similar patterns of response could be identified in simulations with more realistic (i.e., smaller) ocean surface temperature anomalies, once the basic features of the model response for large anomaly simulations had been determined. The results of such experiments are by no means predictions of how the atmosphere itself would respond to such anomalies, although it is gratifying to see some similarities with observed responses. Obviously the model sensitivity study does not exactly reproduce the observed atmospheric response, since in the real climate system the rest of the boundary conditions are not held constant while the ocean surface temperature is changed. Also, there are model shortcomings, such as the poor treatment of cloud

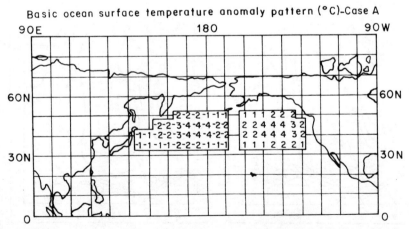

Figure 2.9   The basic ocean surface temperature anomaly pattern (°C) used in the GCM simulations of Kutzbach *et al.* (1977). In one simulation the anomaly was everywhere three times greater than illustrated here

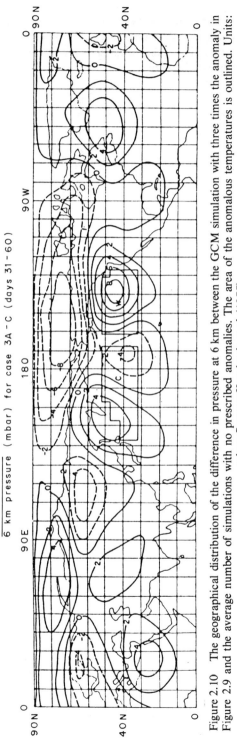

Figure 2.10 The geographical distribution of the difference in pressure at 6 km between the GCM simulation with three times the anomaly in Figure 2.9 and the average number of simulations with no prescribed anomalies. The area of the anomalous temperatures is outlined. Units: mbar. Source: Kutzbach *et al.* (1977)

formation in most models. In the case of the kind of anomaly experiments discussed in this section, the absence of a coupled ocean–atmosphere model is obviously also a shortcoming, since without it the ocean does not respond to changes in the atmospheric circulation.

Nevertheless, it can be seen that both the model and the real atmospheric circulation respond to anomalies in the ocean surface temperature distribution. The fact that models are able to simulate some of the observed large-scale responses also gives confidence in the use of these models for estimating future large-scale effects due to energy systems. Observational and model studies have in addition shown that other anomalies in the natural boundary conditions, such as anomalous areas of surface wetness, or dryness, or anomalous reflectivity over large areas, can similarly influence the large-scale atmospheric circulation. These studies are referred to in particular in Chapter 6.

REFERENCES

Augustsson, T., and V. Ramanathan (1977). A radiative–convective model study of the CO_2 climate problem. *J. Atmos. Sci.*, 448–451.

Ausubel, J. H., and A. Biswas (1980). *Climatic Constraints and Human Activities.* Pergamon Press, Oxford, England.

Barnett, T. P. (1981). Statistical prediction of North American air temperatures from Pacific predictors. *Mon. Wea. Rev.*, **109**, 1021–1041.

Bhumralkar, C. M. (1977). Possible impacts of large solar energy systems on local and mesoscale weather. In, J. Williams, G. Krömer, and J. Weingart (eds.), *Climate and Solar Energy Conversion.* CP-77-9, International Institute for Applied Systems Analysis, Laxenburg, Austria.

Bjerknes, J. (1966). The possible response of the atmospheric Hadley circulation to equatorial anomalies of ocean temperature. *Tellus*, **4**, 820–829.

Bjerknes, J. (1969). Atmospheric teleconnections from the equatorial Pacific. *Mon. Wea. Rev.*, **97**, 163–172.

Bolin, B., and R. J. Charlson (1976). On the role of the tropospheric sulphur cycle in the shortwave radiative climate of the earth. *Ambio*, **5**, 47–54.

Budyko, M. I. (1969). The effect of solar radiation variations on the climate of the earth. *Tellus*, **21**, 611–619.

Budyko, M. I. (1971). *Climate and Life.* Hydrometeorological Publishing House, Leningrad.

Chervin, R. M., and S. H. Schneider (1976a). A study of the response of the NCAR GCM climatological statistics to random perturbations: Estimating noise levels. *J. Atmos. Sci.*, **33**, 391–404.

Chervin, R. M., and S. H. Schneider (1976b). On determining the statistical significance of climate experiments with general circulation models. *J. Atmos. Sci.*, **33**, 405–412.

Chervin, R. M., W. M. Washington, and S. H. Schneider (1976). Testing the statistical significance of the response of the NCAR general circulation model to North Pacific Ocean surface temperature anomalies. *J. Atmos. Sci.*, **33**, 413–423.

Denton, G. H., and W. Karlen (1973). Holocene climatic changes: their pattern and possible cause. *Quat. Res.*, **3**, 155–205.

Flohn, H. (1980). Possible climatic consequences of a man-made global warming. RR-80-30, International Institute for Applied Systems Analysis, Laxenburg, Austria.

Fritts, M. C. (1976). *Tree Rings and Climate*, Academic Press, New York, 567 pp.

GARP (1975). *The Physical Basis of Climate and Climate Modelling.* GARP Publications Series No. 16, WMO-ICSU, World Meteorological Organization, Geneva. 265 pp.

Gates, W. L. (1979). The physical basis of climate. In, *Proceedings of the World Climate Conference.* WMO Publication No. 537, World Meteorological Organization, Geneva.

Hahn, J. (1979). Man-made perturbation of the nitrogen cycle and its possible impact on climate. In W. Bach, J. Pankrath and W. W. Kellogg (eds.), *Man's Impact on Climate,* Elsevier, Amsterdam.

Hanna, S., and F. Gifford (1975). Meteorological effects of energy dissipation at large power parks. *Bull. Amer. Meteor. Soc.,* **56,** 1060–1076.

Huff. F. A., and S. A. Changnon, Jr. (1973). Precipitation modification by major urban areas. *Bull. Amer. Meteor. Soc.,* **54,** 1220–1232.

Jones, P. D., T. M. L. Wigley, and P. M. Kelly (1982). Variations in surface air temperatures: Part 1, Northern Hemisphere, 1881–1980. *Mon. Wea. Rev.,* **110,** 59–70.

Julian, P. R., and R. M. Chervin (1978). A study of the Southern Oscillation and Walker Circulation phenomenon. *Mon. Wea. Rev.,* **106,** 1433–1451.

Kellogg, W. W. (1975). Climatic feedback mechanisms involving the polar regions. In, G. Weller and S. A. Bowling (eds.), *Climate of the Arctic.* Geophysical Institute, University of Alaska, pp. 111–116.

Kellogg, W. W. (1978). Global influences of mankind on the climate. In, J. Gribbin (ed.), *Climatic Change.* Cambridge University Press, Cambridge.

Koenig, L. R., and C. M. Bhumralkar (1974). On possible undesirable atmospheric effects of heat rejection from large electric power centers. R-1628-RC, Rand Corporation, California.

Kukla, G. J., J. K. Angell, J. Korshover, H. Dronia, M. Hoshiai, J. Namias, M. Rodewald, R. Yamamoto, and T. Iwashuna (1977). New data on climatic trends. *Nature,* **270,** 573–580.

Kutzbach, J. E., R. M. Chervin, and D. D. Houghton (1977). Response of the NCAR general circulation model to prescribed changes in ocean surface temperature. Part I. Midlatitude changes. *J. Atmos. Sci.,* **34,** 1200–1213.

LaMarche, V. C., Jr. (1974). Paleoclimatic inferences from long tree-ring records. *Science,* **183,** 1043–1048.

Lamb, H. H. (1969). Climatic fluctuations. In, H. Flohn (ed.), *World Survey of Climatology, General Climatology.* Elsevier, New York, pp. 173–249.

Lamb, H. H. (1972). *Climate: Present, Past and Future. Vol. I, Fundamentals and Climate Now.* Methuen, London, 613 pp.

Lamb, H. H. (1977). *Climate: Present, Past and Future. Vol. II, Climatic History and the Future.* Methuen, London.

Laurmann, J. A., and W. L. Gates (1977). Statistical considerations in the evaluation of climatic experiments with atmospheric general circulation models. *J. Atmos. Sci.,* **34,** 1187–1199.

Leith, C. E. (1973). The standard error of time-average estimates of climatic means. *J. Appl. Meteor.,* **12,** 1066–1069.

Madden, R. A. (1976). Estimates of the natural variability of time-averaged sea-level pressure. *Mon. Wea. Rev.,* **104,** 942–952.

Manabe, S., and R. J. Stouffer (1979). A CO_2-climate sensitivity study with a mathematical model of the global climate. *Nature,* **282,** 491–493.

Manabe, S., and R. J. Stouffer (1980). Sensitivity of a global climate model to an increase of CO_2 concentration in the atmosphere. *J. Geophys. Res.,* **85** 5529–5554.

Manabe, S., and R. T. Wetherald (1967). Thermal equilibrium of the atmosphere with a given distribution of relative humidity. *J. Atmos. Sci.,* **24,** 241–259.

Manabe, S., and R. T. Wetherald (1980). On the distribution of climate change resulting from an increase in CO_2 content of the atmosphere. *J. Atmos. Sci.,* **37,** 3–15.

Namias, J. (1960). Influences of abnormal heat sources and sinks on atmospheric behaviour. *Proc. Intern. Symp. Numerical Weather Prediction, Tokyo.* Meteor. Soc. Japan, 615–629.

Namias, J. (1975). *Short Period Climatic Variations.* University of California, San Diego, 905 pp.

32

Oke, T. R. (1980). Climatic impacts of urbanisation, In, W. Bach, J. Pankrath, and J. Williams (eds.), *Interactions of Energy and Climate*. Reidel, Dordrecht, Holland.

Orlanski, I. (1975). A rational subdivision of scales for atmospheric processes. *Bull. Amer. Meteor. Soc.*, **56**, 527–530.

Rasool, S. I., and S. H. Schneider (1971). Atmospheric carbon dioxide and aerosols: Effects of large increases on global climate. *Science*, **173**, 138–141.

Reck, R. A. (1974). Aerosols in the atmosphere: Calculations of the critical absorption/backscatter ratio. *Science*, **186**, 1034–1035.

Rotty, R. M. (1974). Waste heat disposal from nuclear power plants. NOAA Technical Memorandum ERL ARL-47 US Dept. of Commerce, NOAA, 28 pp.

Rowntree, P. R. (1972). The influence of tropical east Pacific Ocean temperatures on the atmosphere. *Quart. J. Roy. Meteor. Soc.*, **98**, 290–321.

Rowntree, P. R. (1976). Response of the atmosphere to a tropical Atlantic Ocean temperature anomaly. *Quart. J. Roy. Meteor. Soc.*, **102**, 607–625.

Sawyer, J. S. (1965). Notes on the possible physical causes of long-term weather anomalies. WMO Tech. Note No. 66, 227–248. World Meteorological Organization, Geneva.

Schlesinger, M. E., and W. L. Gates (1981). Preliminary analysis of four general circulation model experiments on the role of the ocean in climate. Report No. 25, Climatic Research Institute, Oregon State University, Corvallis, Oregon, USA.

Schneider, S. H., and R. E. Dickson (1974). Climate modeling. *Rev. Geophys. Space Phys.*, **12**, 447–493.

Sellers, W. D. (1969). A global climatic model based on the energy balance of the earth–atmosphere system. *J. Appl. Meteor.*, **8**, 392–400.

Sergin, V.Ya. (1980). A method for estimating climatic fields based on the similarity of seasonal and longer climatic variations. In, J. Ausubel and A. K. Biswas (eds.), *Climatic Constraints and Human Activities*. Pergamon Press, Oxford, England.

Shackleton, N. J., and N. D. Opdyke (1973). Oxygen isotope and paleomagnetic stratigraphy of equatorial Pacific core V28-238: oxygen isotope temperatures and ice volumes on a 10^5 and 10^6 year scale. *Quat. Res.*, **3**, 39–55.

US Committee for GARP (1975). *Understanding Climatic Change*. National Academy of Sciences, Washington, DC, 239 pp.

van der Hammen, T., T. A. Wijmstra, and W. H. Zagwijn (1971). The floral record of the late Cenozoic of Europe. In, K. Turekian (ed.), *The Late Cenozoic Glacial Ages*. Yale University Press, New Haven, Conn., pp. 391–434.

van Loon, H., and J. Williams (1976). The connection between trends of mean temperature and circulation at the surface. Part I. Winter. *Mon. Wea. Rev.*, **104**, 365–380.

van Loon, H., and J. Williams (1977). The connection between trends of mean temperature and circulation at the surface. Part IV. Comparison of the surface changes in the northern hemisphere with the upper air and with the Antarctic in winter. *Mon. Wea. Rev.*, **105**, 636–647.

Walker, J., and P. R. Rowntree (1977). The effect of soil moisture on circulation and rainfall in a tropical model: *Quart. J. Roy. Meteor. Soc.*, **103**, 29–46.

Weiss, R. E., A. P. Wagoner, R. J. Charlson, and N. C. Ahlquist (1977). Sulfate aerosol: Its geographical extent in the midwestern and southern United States. *Science*, **195**, 979–981.

Wigley, T. M. L., M. J. Ingram, and G. Farmer (eds.) (1981). *Climate and History*. Cambridge University Press, Cambridge.

Wigley, T. M. L., P. D. Jones, and P. M. Kelly (1980). Scenario for a warm, high-CO_2 world. *Nature*, **283**, 17–21.

Williams, J. (1979). Modelling the impact of large-scale energy conversion systems on global climate. In, W. Bach, J. Pankrath, and W. W. Kellogg (eds.), *Man's Impact on Climate*. Elsevier, Amsterdam.

Williams, J. (1980). Anomalies in temperature and rainfall during warm Arctic seasons as a guide to the formulation of climate scenarios. *Climatic Change*, 2, 249–266.

Williams, J., and van Loon, H. (1976). The connection between trends of mean temperature and circulation at the surface. Part III: Spring and Autumn. *Mon. Wea. Rev.*, 104, 1591–1596.

CHAPTER 3

Carbon Dioxide

3.1 INTRODUCTION

Since 1958 accurate measurements of the amount of CO_2 in the atmosphere have been made, and, as will be shown in the next section, it is quite clear that the concentration is increasing. It is believed that part or all of this increase is due to the release of CO_2 into the atmosphere by the burning of fossil fuels (coal, oil, and gas). It has also been recently argued that release of CO_2 due to deforestation and soil destruction, especially in the tropics, has contributed to the observed CO_2 increase. Concern arises because of the possibility that substantial increases in the CO_2 concentration would lead to possibly irreversible and undesirable climatic changes. It thus appears necessary to find out whether the increase can be expected to continue, and this involves understanding the reasons for the presently observed increase. It is also necessary to understand the natural 'carbon cycle', that is, the sources and sinks of carbon and the transfers between them. More specifically there are a number of questions that must be answered:

- What has been the rate of addition of CO_2 into the atmosphere from fossil fuel combustion?
- What has been the rate of addition of CO_2 from deforestation and soil destruction? Have other biological processes been simultaneously removing CO_2 from the atmosphere (e.g., enhanced mid-latitude forest growth)?
- Given the total (or net) addition of CO_2 due to this process, how much of it has remained in the atmosphere? What has happened to the CO_2 from fossil-fuel burning, deforestation, and soil destruction that has not remained in the atmosphere?
- Based on our knowledge of the past sources and sinks of CO_2, what will be the future levels of atmospheric CO_2 concentration, given various scenarios for future fossil fuel combustion and deforestation?
- What climatic and environmental changes are likely to be caused by an increased atmospheric CO_2 level?

These questions have received considerable attention in recent publications. Bernard (1980), Singh and Deepak (1980), Kellogg and Schware (1981), Bach *et al.* (1980) and Clark (1982) have considered the CO_2 issue in detail. The following sections review the studies that attempt to answer some or all of the above questions. For further reviews of relevant studies, the reader is referred to the above five books.

3.2 THE OBSERVED INCREASE OF ATMOSPHERIC CO_2 CONCENTRATION

There are many uncertainties regarding the future changes of CO_2 concentration and their effects, but one feature of the CO_2 discussion is clear: the concentration has been increasing steadily during the period since reliable observations began. The long record of the atmospheric CO_2 concentration measured at Mauna Loa in Hawaii (Figure 3.1) shows the mean monthly CO_2 concentration from the beginning of the measurement programme in 1958 until the end of 1978 (Keeling *et al.*, 1976; Slade, 1979). The concentrations are shown by circles connected by a solid line (in two places a dashed line indicates missing data). The average concentration for each year is shown by the horizontal dashed line. The numbers along the bottom show the change of the average concentration from one year to the next. Where data were missing for one or more months, the annual change is within parentheses.

The figure shows two clear variations of the CO_2 concentration: the annual cycle, with a decrease in the Northern Hemisphere summer growing season, and the long-term increase of CO_2 concentration. More detailed studies have suggested that most of the seasonal variation of the atmospheric CO_2 concentration can be accounted for by the land biosphere (Machta, 1979). The annual average concentration of CO_2 measured at Mauna Loa has increased from just over 315 ppm in 1958 to almost 336 ppm in 1978. Measurements at Mauna Loa indicated an atmospheric CO_2 concentration of 338 ppm (by volume) in 1980 (Smith, 1982). Figure 3.2 shows that this increase is confirmed by measurements made elsewhere. Although the amount of the seasonal variation differs from place to place, the long-term increase is clearly global.

It is believed that part, if not all, of this increase is due to burning fossil fuels. Producing cement and flaring natural gas also release CO_2 into the atmosphere, but the amounts are small in comparison to those from burning fossil fuels. The annual values for the global emissions of CO_2 from these sources, calculated for the period 1860–1976 by Rotty (1973, 1977, 1979a), are shown in Figure 3.3a. There is clearly an exponential growth in CO_2 production. Except during the two world wars and the great economic depression of the 1930s, a growth rate of 4.3% per year gives an excellent fit to the data. Figure 3.3b shows the annual global production of CO_2 from fossil fuels and cement for the period 1950–1980 (Rotty, 1981). It is clear that since 1973 there has been a slowing in the global rate of CO_2 production.

During the past 20 years 1–2 ppm CO_2 were added to the atmosphere each year by burning fossil fuels, and to a much lesser extent by cement production and natural gas flaring. Figure 3.1 shows that the annual increase in atmospheric CO_2

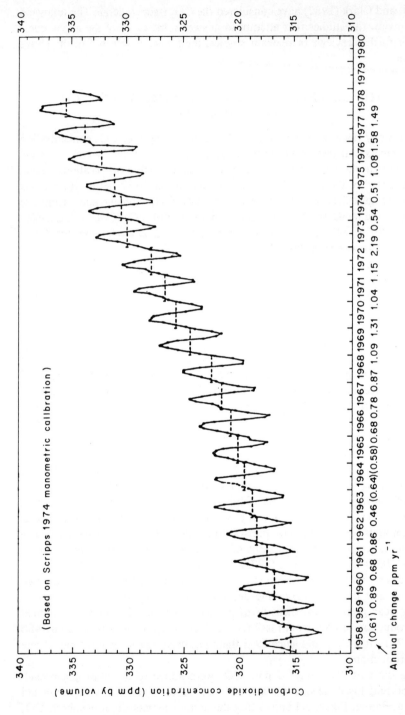

Figure 3.1 The concentration of atmospheric CO_2 at Mauna Loa Observatory, Hawaii. Mean monthly concentrations are shown as circles connected by a solid line; a dashed line between data indicates missing data. The horizontal dashed lines locate mean annual concentrations. The numbers at the bottom of the figure show the change of annual concentration. The parentheses indicate the absence of data for one or more months during the year for which the average was derived. Source: USDOE (1979)

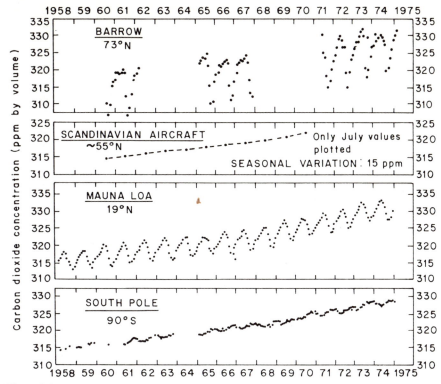

Figure 3.2 The growth of atmospheric CO_2 concentration at four locations. Except for the Scandinavian aircraft data, the points denote monthly average concentrations. In Scandinavia the values have been referred to July, using an average pattern of seasonal variability. Source: Machta (1979)

concentration ranges between about 0.5 ppm and 1.5 ppm. Therefore, as a very rough estimate, the annual atmospheric increase is about half of the fossil-fuel input. If it is assumed that fossil fuels (and cement and natural gas flares) are the only sources of CO_2, then one can calculate the per cent of the fossil-fuel CO_2 which remains airborne each year, usually referred to as the 'airborne fraction', which is shown in Figure 3.4. It is quite variable, ranging between 25% and more than 100%. Machta (1979) has pointed out that if the annual increase of atmospheric CO_2 were calculated from more stations than just Mauna Loa, the airborne fraction would not vary so much from year to year. In addition, studies have suggested that large-scale tropical atmosphere–ocean interactions also modulate the atmospheric CO_2 concentration and this affects the airborne fraction (e.g., Bacastow, 1976). The long-term airborne fraction from 1959 to 1974 based on the average of the observations at Mauna Loa and at the South Pole was calculated by Bacastow and Keeling (1979) to be 56.5%; they pointed out that this fraction is uncertain, owing in particular to systematic errors, of the order of 15%, in the fossil-fuel consumption data.

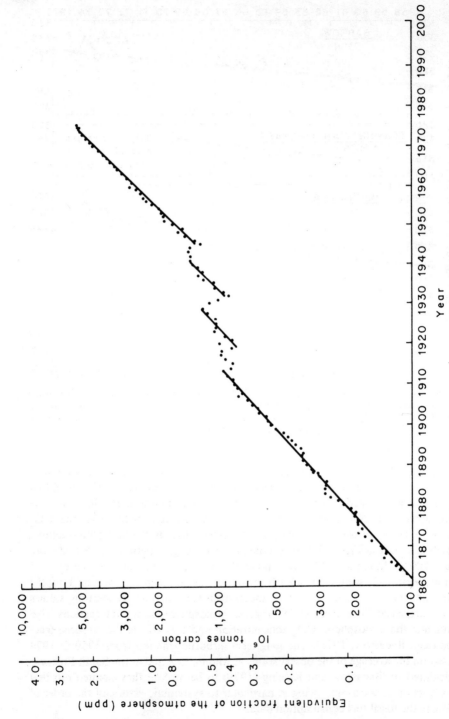

Figure 3.3a   Carbon dioxide production from fossil fuels and cement. Source: Rotty (1979a)

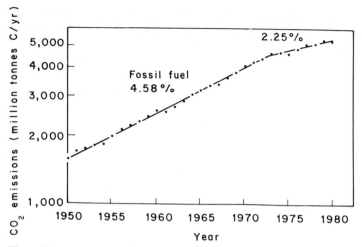

Figure 3.3b   Annual global production of CO_2 from fossil fuel and cement with indicated growth rates. Source: Rotty (1982)

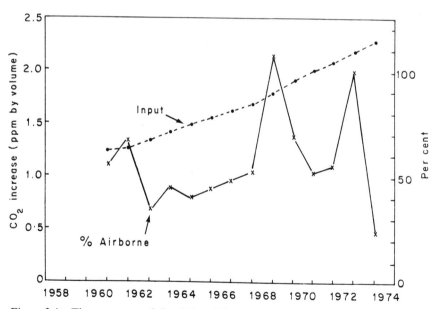

Figure 3.4   The per cent of fossil-fuel CO_2 remaining airborne each year (lower curve), assuming that the observed Mauna Loa CO_2 growth represents global conditions. The upper curve shows the amounts of increase expected from fossil fuels assuming that the input CO_2 mixes rapidly with the entire atmosphere. Source: Machta (1979)

Very few data are available for the period before regular accurate observations of atmospheric CO_2 concentration began in 1957. It has been estimated that the preindustrial concentration of CO_2 was 265–290 ppm (Bolin, 1979) or 285–305 ppm (Machta, 1979). Generally a value of 290–295 ppm is assumed for the middle of the 19th century. Callender (1958) obtained an average of 290 ppm (by volume) from measurements in West Europe between 1870 and 1900. Southern Hemisphere data from the 19th century suggest a lower atmospheric CO_2 concentration of about 270 ppm (by volume) (results by Wigley, cited by Smith, 1982).

Figures given above suggest that roughly 50–60% of the CO_2 released each year by fossil-fuel combustion remains in the atmosphere. As far as predictions of the future concentration of CO_2 in the atmosphere are concerned, the crudest estimate would be that in the future about 50% of the CO_2 released by fossil fuels will continue to remain airborne each year. This can be, and has been, challenged on several grounds. Firstly, it is necessary to know where the 40–50% of the CO_2 not remaining in the atmosphere goes to. The possible sinks are the ocean and the biosphere. Secondly, it cannot be assumed that fossil fuels (and cement manufacture and natural gas flaring) are the only man-made sources of atmospheric CO_2. As will be discussed in much more detail in subsequent sections, it has been suggested (e.g., Woodwell, 1978) that a further man-made source of atmospheric CO_2 is deforestation, especially in the tropics, and soil deterioration. As an extreme case, if it is assumed that these sources add just as much CO_2 to the atmosphere as the fossil fuel burning, then the real airborne fraction would be half as large as that when biospheric sources are ignored. This suggests that, in order to make reliable predictions of the future concentrations of atmospheric CO_2, it is necessary to have an understanding of the natural and anthropogenic sources and sinks of CO_2. On the global scale, the reservoirs of carbon (atmosphere, ocean, biosphere, lithosphere) and the transfers between them are the components of the 'carbon cycle'.

3.3 THE CARBON CYCLE

Our present knowledge of all aspects of the carbon cycle has been reviewed comprehensively by Bolin et al. (1979); Figure 3.5, based on their study, shows its major components. The four carbon reservoirs are the atmosphere, the ocean, the biota, and sediments (including fossil-fuel deposits). The reservoir sizes and the fluxes are uncertain and some fluxes are not well enough known to be included in the schematic illustration of the carbon cycle.

In 1980 the atmosphere contained about 710×10^{15} g carbon, which can be compared with 610×10^{15} g carbon estimated for 1860. The ocean can be divided into three layers: the surface mixed layer with about 680×10^{15} g carbon; the intermediate layer with about $8,200 \times 10^{15}$ g carbon; and the deep layer with about $26,000 \times 10^{15}$ g carbon.

The deep ocean exchanges only very slowly with the other ocean layers and thus the water is found to be of the order of 1,000 years old.

Estimates of the amount of carbon in living matter on land vary. Machta and Elliot (1980) suggest, as shown in Figure 3.5, a living matter content of 590×10^{15} g carbon. Bolin (1979) suggests that about 800×10^{15} g carbon are stored on land as living organic matter. Likewise, estimates of the total amount of carbon in the soil vary between $1,000 \times 10^{15}$ g carbon and $3,000 \times 10^{15}$ g carbon. The uncertainty largely arises because of different values assumed for the carbon stored as peat, but, since the transit time for carbon through this pool is of the order of thousands of years, the uncertainty is not so important when perturbations due to man's activities of the order of hundreds of years or less are considered. The amount of carbon in recoverable fossil fuels (90% in coal) is generally given as $5,000 \times 10^{15}$ g (Bolin, 1979; Munn and Machta, 1979), although estimates are variable.

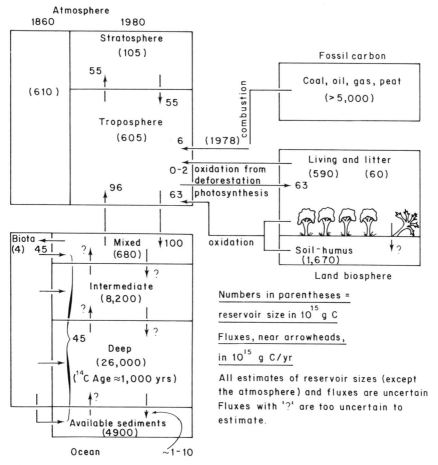

Figure 3.5   The global carbon cycle. Source: Machta and Elliot (1980), based on Bolin *et al.* (1979)

Numerical models of the carbon cycle are used to predict the CO_2 concentration. One of the most common model types is the box model, which considers the reservoirs of carbon and the transfers of carbon between them (e.g., Keeling and Bacastow, 1977). Figure 3.6 shows a six-reservoir model. In this case the atmosphere, oceans, and land biota are each divided into two boxes. The stratosphere and the short-cycled biota exchange carbon extremely rapidly with the atmosphere. In both cases, the total carbon in the reservoir is changed once in less than 25 years. Therefore these reservoirs have very little effect on the storage of anthropogenic CO_2 and the model can generally be reduced to four reservoirs with only the oceans subdivided. There are various ways that the models can be tested to see if they simulate the carbon cycle realistically (Broecker, 1977). For example, a model should be able to reproduce the observed increase in atmospheric CO_2 content between 1958 and 1978. It should also be able to simulate the decrease, since 1963, of bomb-produced radiocarbon ^{14}C.

There are, according to the description of the carbon cycle in Figure 3.5, two possible sinks for the CO_2 released into the atmosphere by fossil-fuel combustion: the ocean and the land biota. The capacity of the global oceans to store carbon is tremendous, but the transfer from the atmosphere to the deep ocean is slow except, perhaps, in the high latitudes. The physical–chemical processes that transfer carbon from the atmosphere to the surface ocean are quite well understood. There is a so-called 'buffering action' by the ocean due to the pre-

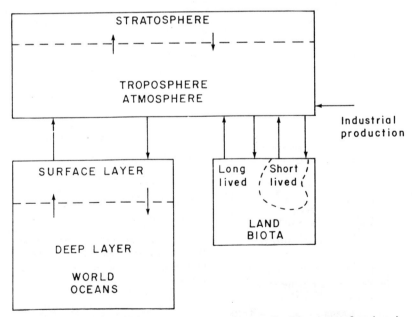

Figure 3.6 A six-reservoir model of the carbon cycle. The mass of carbon in each reservoir is represented by $N_j(t)$ and the transfer coefficients between reservoirs are given by l_j and k_j. Source: Bacastow and Keeling (1973)

sence of carbonate, bicarbonate, and borate ions in the sea water. If the atmospheric CO_2 content increases by $x\%$, the resulting relative increase of surface oceanic CO_2 will, in equilibrium, only be $x/\xi\%$. The value of the buffer factor ξ at the present time is believed to be about 10 (e.g., Keeling and Bacastow, 1977). It is possible that the buffer factor would increase as further CO_2 is released by fossil-fuel combustion.

The land biota can act either as a source or as a sink of carbon. The main storage is in the forests, especially in the tropics and the Northern Hemisphere temperate and boreal forests. Clearing forests has been claimed to be a source of atmospheric CO_2 (e.g., Woodwell, 1978, and Section 3.4(b)). Likewise the deterioration of soils, due, for instance, to bad farming practices, also releases CO_2. On the other hand, increased atmospheric CO_2 could stimulate the plants, giving an increase in the mass of the biota which would thereby be acting as a sink. This process is referred to as CO_2 fertilization, and is seen to operate in the case of greenhouses where growth is stimulated by increasing the CO_2 content of the air. Outside of the greenhouse, where water and nutrient supplies are sometimes limiting, it has been assumed that a 10% increase in atmospheric CO_2 concentration would lead to a 0–3% increase in the biomass and this fertilization effect has been taken into account in carbon cycle models (e.g., Oeschger et al., 1975; Keeling, 1973). At present, however, most ecologists believe that the amount of CO_2 fertilization must be small because of the limitations imposed by the availability of water and nutrients and the ease with which respiration and the decay of organic matter are stimulated by disturbance (Woodwell, 1980).

Bacastow and Keeling (1979) have made a series of sensitivity tests of the six-reservoir model shown in Figure 3.6. Firstly, the response to four different inputs of fossil-fuel CO_2 was investigated. The inputs are shown in Figure 3.7. In each case, the production increases to a peak and declines, the higher the peak the faster is the subsequent decline in production. Figure 3.8 shows the atmospheric CO_2 concentration predicted by the six-reservoir model for the CO_2 production curves in Figure 3.7. The CO_2 increases to a peak between the years 2100 and 2300 and declines slowly thereafter. The peak level is delayed if the period of CO_2 production is stretched out. The peak value is only slightly reduced when the production is slower. In the extreme input curves, the maximum production rate differs by 327%, but the resulting peak atmospheric CO_2 concentrations differ only by 15% (Bacastow and Keeling, 1979). In both cases the CO_2 input is a rapid impulse in comparison to the long transfer time of CO_2 into the subsurface ocean. The slow decrease of atmospheric CO_2 is also due to the slow transfer into the ocean.

A second set of simulations using the six-reservoir model looked at the influence of assumptions regarding the land biota. Figure 3.9 shows the results. In the simulation which produced the lowest curve the biota were allowed to grow without limit owing to CO_2 fertilization. The biota carbon almost tripled in size and the atmospheric CO_2 was reduced to approximately the preindustrial level within 300 years. Other simulations assumed an arbitrary cut-off of biota growth. The upper curve results from the cut-off of biota growth in the year 2025. The

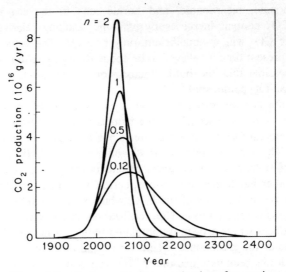

Figure 3.7 Carbon dioxide production for various assumed patterns of fossil-fuel consumption. The ultimate production of CO_2 is fixed at 8.2 times the amount of CO_2 in the preindustrial atmosphere. Source: Bacastow and Keeling (1979)

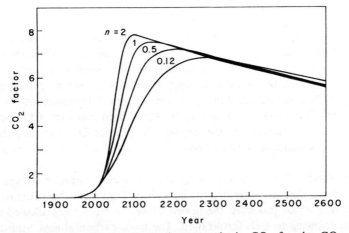

Figure 3.8 Predicted increase in atmospheric CO_2 for the CO_2 production patterns shown in Figure 3.7. The CO_2 factor is the ratio of the amount of CO_2 in the atmosphere to the amount in the pre-industrial atmosphere. Source: Bacastow and Keeling (1979)

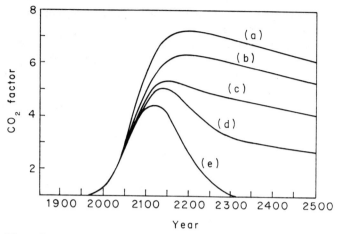

Figure 3.9   Effect of biota growth on the predicted increase in atmospheric CO_2: (a) Standard case, same as the $n = 0.5$ curve in Figure 3.8. The biota growth factor was set linearly to zero between the years 2000 and 2025. The ultimate biota increase is 6% relative to its preindustrial value. (b) Biota growth factor set linearly to zero between the years 2000 and 2150. Ultimate biota increase is 42%. (c) Biota growth factor set linearly to zero between the years 2000 and 2250. Ultimate biota increase in 98%. (d) Biota growth factor set linearly to zero between the years 2000 and 2350. Ultimate biota increase is 165%. (e) Biota grow without a cut-off. The biota carbon pool continues to increase until all of the fossil fuel input is in the biota. Ultimate biota increase is 308%. Source: Bacastow and Keeling (1979)

biota carbon pool increases in this case by 6% compared to the preindustrial size. Bacastow and Keeling (1979) concluded that as long as the biota size does not more than double, the atmospheric CO_2 concentrations are similar to the basic model results discussed above except lowered by the amount the biota takes up. Larger increases in the biota succeed in removing the fossil fuel CO_2 quickly from the atmosphere but such large increases are unrealistic in the real world, especially in view of the increasing population's demand for land and food.

A second type of model has been developed by Oeschger et al. (1975), in which the ocean is represented by a wind-mixed layer 75 m deep coupled with a deep ocean in which transport is by vertical eddy diffusion. The vertical diffusivity is determined so that the model correctly simulates the distribution of radiocarbon. This 'box-diffusion' model is therefore slightly more plausible on physical grounds than a box model in which there is no smooth gradient of CO_2 or radiocarbon. However, the box-diffusion model is still unrealistic, since horizontal motions and vertical motions are probably just as important as vertical diffusion for CO_2 transport and these are not included in the model (Bacastow and Keeling, 1979).

Bacastow and Keeling (1979) have compared the results of the box-diffusion and box model, both adjusted to fit the same steady-state radiocarbon data and the atmospheric CO_2 increase from 1959 to 1974, and find that both models give similar predictions over the next 1,500 years.

Killough and Emanuel (1981) have compared five models of carbon circulation in the global ocean. They were required to be consistent, as far as possible, with the hypothesized depth distribution of radiocarbon activity. The five models divided the ocean into a surface layer and a deeper portion. The first two considered a surface layer and a deep-water reservoir that exchanged carbon by linear fluxes. The three remaining models consisted of a 75-m deep surface layer and a deep reservoir subdivided into 18 horizontal layers. The layered models differed from each other in the topological configuration of fluxes among the layers.

These authors found that the five models exhibited significantly different responses to fossil CO_2 and weapons ^{14}C regimes. For instance, of the estimated 134×10^{15} g of fossil fuel carbon produced in the period 1860–1975, the models estimated a net uptake of 24×10^{15} g carbon to 66×10^{15} g carbon by the ocean.

Killough and Emanuel conclude that empirically calibrated box models offer a flexible and powerful tool for exploring a variety of hypotheses relating to carbon cycle dynamics. They believe, however, that ultimately general circulation models that incorporate detailed patterns of water transport, air–sea gas exchange, and improved quantification of dynamic chemical and biological processes will become the chosen research tools.

Most of the models of the carbon cycle used to predict future atmospheric CO_2 concentrations are one-dimensional in character. Machta and Elliot (1980) have pointed out that more realistic models should consider the ocean and biosphere in more detail. Moreover the same authors state: 'The limitations on modeling are not now in computer hardware but in understanding the system and in data inadequacies.'

Most of the models so far have not considered the difference between warm and cold ocean areas. In fact the warm ocean waters of the equatorial belt are supersaturated with respect to atmospheric CO_2 and thus act as a CO_2 source for the air. The cold ocean waters in high latitudes are undersaturated and thus act as a CO_2 sink. The atmospheric CO_2 concentration is the result of the balance between CO_2 fluxes from sources (equatorial oceans, biological decay of organic debris, volcanic exhalations, fossil fuel combustion) and absorption into sinks (polar oceans, plant photosynthesis, and organic debris accumulation). According to Machta and Elliot (1980) the oceans exchange $90–100 \times 10^{15}$ g carbon as CO_2 each year, with a net uptake in high latitudes and a net release in the warm tropics. The net release from the tropical ocean might be a significant fraction of the amount released from the combustion of fossil fuels and therefore it is understandable that relatively minor fluctuations of equatorial ocean temperatures appear to be observable in the record of atmospheric CO_2.

The carbon cycle of the earth is very complex, with a number of reservoirs of carbon and chemical, physical, and biological processes linking them. Numerical models of the carbon cycle are necessary so that predictions of the future concentrations of atmospheric CO_2 can be made. Carbon cycle modelling was discussed by a working group at the IIASA Workshop on Carbon Dioxide, Climate and Society (Williams, 1978). This group pointed out that, although there are many uncertainties in models of the carbon cycle (more uncertainties than there were 10 years ago), there are still possibilities for some prediction of future atmospheric CO_2 concentrations. Past fossil fuel injection rates are known quite well and the observations of the atmospheric CO_2 concentration made during the past 20 years indicate that about 50% of the fossil fuel CO_2 has remained in the atmosphere. Since it is unlikely that biogeochemical conditions will change significantly during the next 20–30 years, predictions over this period should be possible using extrapolation, given an approximately exponential growth rate in fossil fuel use. The existing models do not disagree over this time scale. However, beyond the next 30 years the uncertainties in predictions increase rapidly.

As suggested above, it is the role of the biosphere that is most uncertain. The working group felt that there is fair agreement on the role of the ocean, i.e., on how much CO_2 the oceans can or cannot partition from the atmosphere. It was concluded that forest fires and a number of changing land-use practices were not significant sources of CO_2, but that tropical forest clearing could represent a significant source. It was emphasized that it is extremely difficult to use the data of observed tropical forest clearing to derive an estimate of the global inputs of CO_2, and this fact is reflected in order-of-magnitude range of estimates in recent literature. Given the growing consensus that there probably has been net global deforestation, it was concluded that this may have been partly compensated by regrowth patterns in areas cut over past decades. This is confirmed by subsequent studies reported in Section 3.4(b).

3.4 FUTURE ANTHROPOGENIC INPUT OF CO_2 INTO THE ATMOSPHERE

As indicated earlier in the chapter, there appear to be two main man-made sources of atmospheric CO_2: the combustion of fossil fuels; and deforestation/soil deterioration. An accurate prediction of the future rates of input is required, in order that forecasts may be made of potential climate changes, etc. This section will consider past input rates and predictions.

(a) Carbon dioxide additions to the atmosphere from fossil-fuel combustion

The annual values for the global emission of CO_2 from the combustion of fossil fuels and (to a much lesser extent) from producing cement and flaring natural gas were shown in Figure 3.3. A growth rate of 4.3% per year fits the data, except during the two world wars and the years of the Great Depression. Two questions first come to mind with regard to predictions of future rates of fossil fuel use: Will fossil fuel use continue to grow? How long will the fossil fuels last?

As pointed out in the IIASA energy study (Energy Systems Program, 1981), the global commercial primary energy use in 1975 totalled about 8.21 TW. Of this 2.36 TW was coal, 3.81 TW was oil and 1.42 TW was gas. Figure 3.10 shows the

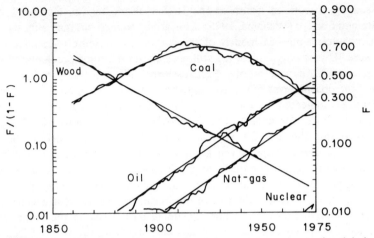

Figure 3.10   Historic contribution of various energy sources to the global energy supply. The smooth line is computed and the irregular line represents the observed data. Source: Energy Systems Program (1981)

contributions of various energy sources to the global energy supply from 1860 to 1975. We see, first of all, that wood was the major energy source in 1860, with about 70% of the market share. Wood was then steadily replaced by fossil fuels. The use of coal peaked in the early 1900s and was replaced by oil and natural gas. One further feature of Figure 3.10, which could be of significance to the CO_2 question, is that substitution by new energy sources is not a fast process and is constrained by market forces, which in the past have led to an extremely regular (and predictable) pattern of substitution.

Based on the information shown in Figure 3.10, it is possible to explore the market penetration constraints affecting primary energy technologies in future energy systems. The IIASA Energy Systems Program (1981) shows, for example, that, if it is assumed that nuclear power would begin to penetrate the primary energy market in 1979 at a commercially significant rate of 2 to 3%, then it would have increased its share to 40% by the year 2030. The market penetration curves also indicate that natural gas would take on an ever-increasing share, reaching a peak of almost 60% around 2015 and declining thereafter. Since natural gas releases less CO_2 than coal does per unit of energy converted, both of the above projections based on the market penetration curves imply a low CO_2 release during the next 50 years.

The IIASA Energy Systems Program (1981) estimates that globally the conventional fossil fuel resources are of the order of 1,000 TWyr. However, they also point out that the exploitation of unconventional fossil fuels (e.g., tertiary oil recovery, heavy oils, tar sands, shale oil) would bring the global fossil fuel resources to about 3,000 TWyr (here 'resources' refers to deposits that are known fairly well or exist as estimates of what we might find if we look harder). The cumulative primary energy consumption by the year 2030 was estimated by this

IIASA study to be of the order 900–1,400 TWyr, and this gave rise to the suggestion that the fossil fuels could be a bridge between today and the long-term future era of sustainable non-fossil energy.

Figure 3.11 synthesizes various considerations made by the IIASA study with regard to future fossil fuel use. Firstly, the curves suggest that an early decline of production levels must not necessarily be expected. It is also suggested that discovery rates could possibly be maintained—at a cost—at a more or less continuous level for a few more decades. The production levels in 2030, according to Figure 3.11, could be more than twice the production levels in 1975.

Table 3.1 shows two scenarios developed by the IIASA Energy Systems Program (1981) for the year 2030. In the high scenario the global primary energy supply total is 4.3 times higher than that of 1975, while in the low scenario it is 2.7 times higher. The use of fossil fuels is 2–3 times larger in 2030 than in 1975. The implications for this increase on the atmospheric CO_2 concentration are discussed in Section 3.4(c). In both of the scenarios coal is used not only in solid form, but also the demand for gaseous and liquid fuels is seen to necessitate coal use in synthetic fuels production. An analysis of the impact of CO_2 production from synthetic fuels is also included in Section 3.4(c).

Many studies, some of which are referred to subsequently, have made projections of the future use of fossil fuels and the consequent atmospheric CO_2 release. Such studies generally make a prediction of the energy demand in the future and assumptions about how much of this demand will be satisfied by fossil fuels. Ausubel (1980) has looked at the problem differently, by investigating how much fossil fuel would have to be burned to give a certain CO_2 concentration by a given

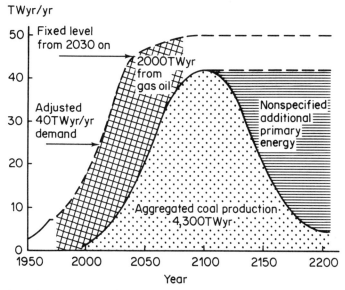

Figure 3.11 A theoretical curve showing the potential levelling of fossil fuel production. Source: Energy Systems Program (1981)

Table 3.1 The use of fossil fuels in two supply scenarios of global primary energy, 1975–2030 (TWyr/yr). Source: Energy Systems Program (1981)

Primary source	1975	High scenario 2000	High scenario 2030	Low scenario 2000	Low scenario 2030
Oil	3.62	5.89	6.83	4.75	5.02
Gas	1.51	3.11	5.97	2.53	3.47
Coal	2.26	4.94	11.98	3.92	6.45
Total (fossil fuels and other sources)	8.21	16.84	35.65	13.59	22.39

time. A comparison of the two approaches, as made by Ausubel, is instructive for exploring what kind of world would accompany high levels of CO_2.

As an example of the first approach, Ausubel (1980) cites the study of Rotty (1978). Figure 3.12 shows the global production of CO_2 by world segments in 1974. The developing countries produced only 13% of the CO_2 at that time. Rotty (1978) made a projection of the CO_2 production in 2025, on the basis of predicted population growth rates and improved living standards in the developing countries. Figure 3.13 shows the predicted global production of CO_2 by world segments in the year 2025, and the 1974 values are shown to the same scale. The

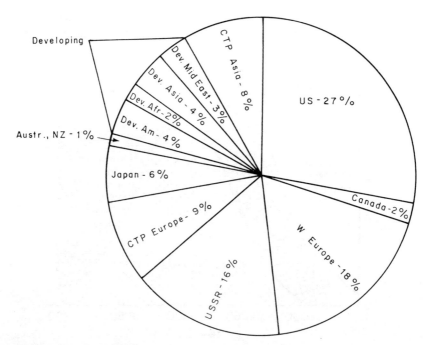

Figure 3.12 Global CO_2 production by world segments, 1974. Source: Rotty (1978)

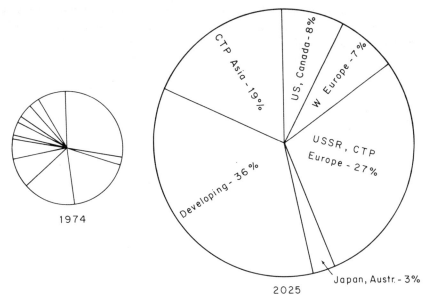

Figure 3.13   Global CO_2 production by world segments predicted for 2025. The 1974 production is shown to the same scale for comparison. Source: Rotty (1978)

total CO_2 production in 2025 is estimated to be 4.5 times the 1974 amount, i.e., it is greater than that in the high scenario of Table 3.1. Rotty's projection suggests that the CO_2 production of the developing countries will be 40% of the total in 2025. Thus, significant responsibility for the CO_2 increase is placed on the developing countries.

On the other hand, Ausubel (1980), using current estimates of resources, finds that the carbon that could have major impacts within a time scale of 50–100 years would be extracted by a small number of countries, which with one major exception are advanced industrialized countries. This leads to the conclusion that for CO_2 to be considered a truly global issue, one has to examine the *trade* in carbon and the consequences of climatic change, as well as energy demand and production.

As a working hypothesis, Ausubel assumes that, for climatic change to be an urgent problem, a cumulative production of 1,500 gigatonnes (Gt) of carbon would be required by the year 2050. Ausubel then analyses the contributions of carbon from man-made destruction of the biota. It is roughly estimated that 75 Gt of carbon were released to the atmosphere by forest clearing in the period 1860–1970 (Revelle and Munk, 1977). More than 30 Gt of this may have come from North America, Europe, the USSR, Australia and New Zealand; that is, the contributions from industrialized and developing countries were not greatly different. The most important result is that the biota contribution is small compared with the assumed 'requirement' of 1,500 Gt by the year 2050. Ausubel further finds that, in view of the historic contributions and presently available biospheric carbon stocks, the maximum feasible exploitation would give 10–20% of the CO_2 amount 'required', if vast combustion of forests and peat is allowed. Such large-scale deforestation would have serious environmental and societal impacts, but would not contribute greatly to the CO_2 problem.

If the working-hypothesis limit of 1,500 Gt of carbon is to be reached in the year 2050, it would mainly have to come from fossil fuels. It has been estimated by Bolin (1979) that a cumulative total of about 140 Gt of carbon has been injected into the atmosphere by fossil fuel combustion; 80% of this is estimated to have come from the industrialized world. Ausubel (1980) finds that the historic fossil fuel contribution of about 140 Gt carbon, plus exploitation of all fossil fuel reserves (that is, known stocks that could be produced economically today) of about 540 Gt carbon, and the maximum biotic contribution of about 240 Gt carbon would only give about two-thirds (61%) of the level of 1,500 Gt taken as a working hypothesis. Secondly, Ausubel finds that the sum of all conventional oil and gas reserves and resources is currently estimated at less than 450 Gt carbon, which suggests that oil and gas are not likely to be the major contributors to any CO_2 problem. It is shown that CO_2 would be a serious problem mainly with long-term, large-scale development of coal resources by a small number of countries.

There are two reasons Ausubel cites suggesting that such large-scale development of coal would be very difficult. Firstly, to extract coal at the required rate would not be easy. The IIASA Energy Systems Program (1981) describes the difficulties:

> [For North America] a ceiling of nearly 2.9 *billion tons per year* (2,700 GWyr/yr) of coal production has been assumed—based on limitations of rail transport, water availability for Western US surface mining, limited coal uses (as coal) within the US and Canada, few deep water port facilities, and air quality and other environmental considerations . . . Three billion tons of coal per year in the US is an enormous mining activity; it is nearly five times the 1978 US production rate. . . . It is nearly twice the present coal mining of the whole world. It means great capital investments, large land uses, substantial numbers of new miners. Presumably it could be done. . .

The second difficulty with large-scale exploitation is the need for international trade. Coal resources and reserves are very unevenly distributed worldwide. The nations with large coal resources are China, the USSR, and the USA. These countries are estimated to have close to 90% of the global resources. For many nations to pursue a coal-energy strategy, a large coal trade would be required; the IIASA Energy Systems Program (1981) has indicated that this could be problematic:

> One major question relates to the possible supply role of the three coal giants (the USSR, the US, and China) which have tremendous

resources but also very large domestic energy requirements. It is not clear whether these countries would be willing and/or able to manage additional capacity for an export market, although this may serve political or economical purposes. Potential customers could also be reluctant to increase their energy dependence on these political giants.

Thus, in contrast to the approach taken, for example, by Rotty (1978), Ausubel (1980) suggests that the CO_2 problem is not likely to have a large contribution from developing countries unless there is enormous coal trade from the US, the USSR, and China. Constraints on international coal trade and the technical feasibility of large-scale coal exploitation could, according to Ausubel, make a doubling of atmospheric CO_2 difficult to reach in the next 100 years, and higher levels look very remote from this perspective.

(b) Carbon dioxide inputs to the atmosphere from biotic sources

In Section 3.3 the observed annual increase in atmospheric CO_2 concentration was shown to be 48–56% of the annual production of CO_2 from fossil fuels. Oceanographic studies have shown that the ocean is capable of absorbing 35–40% of the fossil fuel CO_2 (e.g., Oeschger et al., 1975) and it seemed that the budget balanced to within a few per cent, with the land biota being assumed to absorb the remainder of the fossil fuel CO_2. On the basis of this information, several forecasts of the future evolution of the atmospheric CO_2 content were made (e.g., Machta, 1973; Niehaus, 1976; Keeling and Bacastow, 1977).

Recently, however, ecologists challenged this view by proposing that there are large additional inputs of CO_2 to the atmosphere through the destruction of vegetation and soil matter, mainly in the tropics (e.g., Woodwell, 1978; Woodwell et al., 1978). If the input of CO_2 from deforestation, etc., is of the same order of magnitude as the fossil fuel CO_2 input, then the mechanisms for removing CO_2 from the atmosphere must be correspondingly more efficient. Oceanographic studies, however, have continued to show that a significantly higher uptake of CO_2 by the ocean is incompatible with established knowledge of carbon chemistry and vertical mixing in the ocean (e.g., Siegenthaler and Oeschger, 1978; Björkström, 1979; Broecker et al., 1979).

The tropical deforestation and soil destruction result from demographic, political, and economical pressures. Estimates of the rate at which closed forests are disappearing range from 10 to 20 million hectares per year; nearly all of this occurring in the humid tropics. Of the total wood used in developing countries 80% is burned for fuel and much of it never passes through a commercial market. However, the spread of agriculture appears to be the major cause of deforestation (Eckholm, 1979). Knowledge about the extent of forests or the rate of deforestation is incomplete because only about half of the world's forests have been subjected to detailed surveys and only a fraction have been surveyed more than once, so that changes over time cannot yet be calculated. Even though improved aircraft and satellite monitoring could give more reliable estimates, there will still be uncertainties because, for example, clearance is incomplete in many cases so that man's activities often result in a continuous thinning and degradation of the vegetation, which is imperceptible on a short time scale (Hampicke, 1979). A recent workshop (Woodwell, 1980) discussed the possibility of measuring changes in the area of forests globally through the use of satellite remote sensing. It was concluded that a programme based on LANDSAT imagery supplemented by aerial photography would be possible and appropriate. LANDSAT (or Earth Resources Technology Satellite Program) was established for the purpose of using spacecraft to monitor resources with a scanning system on a global scale. With LANDSAT is is possible to identify surface features of the earth on a large scale, and therefore it could be used to assess the pools and fluxes of carbon.

Clearing forests for agriculture is accompanied by a decline in the amount of soil carbon (Woodwell, 1980). For example, soils in the tropical rainforest may have a carbon content of 2–5%, and this may fall by 50% or more under agricultural use. Similarly, soils in the savannah zone may also lose half of their carbon on conversion to agriculture. These changes are caused by loss of vegetation, by increased temperatures and increased rates of respiration, and, in some cases, by erosion (Woodwell, 1980). In the temperate zone, there is a similar but smaller change in carbon content. In some places, especially in North America, reversion of farmland through succession to woodland and forest is taking place. This may be accompanied by a build-up of soil carbon, but the magnitude of the increase is not known (Woodwell, 1980).

Hampicke (1979) calculates that the annual net release of carbon due to tropical deforestation and soil deterioration in 1977 amounted to $1.7–3.9 \times 10^{15}$ g carbon. Bolin (1979) calculates that the annual input, primarily due to tropical deforestation, is $1–5 \times 10^{15}$ g carbon. Woodwell et al. (1978) compiled several estimates of the annual release of carbon from the world's biota. The range of estimates is large: from -1×10^{15} g (net increase of the biota), to a maximum of 18×10^{15} g carbon, with a most likely range of $4–8 \times 10^{15}$ g carbon. Baumgartner (1979) suggests that the net effect of changes in global forest cover is a release of 3.3×10^{15} g carbon per year. Therefore, there seems to be a consensus that the biota is a source of atmospheric CO_2 at the present day, and in many cases the input from this source is estimated to be of the same order of magnitude as the fossil fuel CO_2 input to the atmosphere.

Hampicke (1979) has investigated potential sinks for atmospheric CO_2 in the land biota and estimates that the annual removal by natural accretion in soils, river discharge, biomass increase in recovering temperate forests and due to CO_2 fertilization is about $0.5–2.8 \times 10^{15}$ g carbon. Thus Hampicke finds that it is difficult, from an ecological point of view, to conclude that the exchange of carbon between the atmosphere and land ecosystems is balanced and it seems likely that the additions of CO_2 to the atmosphere through man's interference exceed the removal.

Most reviews of the role of deforestation in the carbon budget have concentrated on the tropical forests, because of their large biomass (53% of the total forest biomass) and the continuing reduction of tropical forest area and volume

(Armentano and Hett, 1980). The second largest biotic carbon pool is the temperate zone forest, most of which is between 30°N and 60°N. In these forests most of the forest destruction occurred in the past—50 to 200 years ago in North America and 2,000 to 5,000 years ago in the Far East (Armentano and Hett, 1980). A recent workshop—Carbon Balance in Northern Ecosystems and the Potential Effect of Carbon-Dioxide-Induced Climatic Change (Miller, 1980)—found that forest inventory data from most of the commercial timberlands of the northern temperate zone suggested that these forests have acted as an annual sink for about 10 Gt carbon during the past several decades (Armentano and Hett, 1980. Another workshop—The Role of Organic Soils in the World Carbon Cycle: Problem Analysis and Research Needs (Armentano, 1980)—concluded that organic soils that are being drained are releasing a small amount of carbon, while a smaller amount is being stored in undrained soils. It is also suggested (see Machta and Elliot, 1980) that freshwater lakes and streams may be storing an additional quantity equal to about 5% of the fossil-fuel emission.

The role of the tropical forest has been discussed in detail at a recent symposium in Puerto Rico (Brown et al., 1980). Lugo (1980) examined the question of whether tropical forest ecosystems are sources or sinks of carbon and concluded that even deforestation could create carbon sinks if succession is fast enough and if a portion of the initial biomass remains on site. He suggested that climax ecosystems may also be sinks of carbon through export of organic matter to downstream aquatic ecosystems. Brown and Lugo (1980) estimated the organic carbon storage in tropical forests and found a lower value than earlier estimates. The estimate of Whittaker and Likens (1973) is almost double that of Brown and Lugo (1980). The difference is attributed to differences in definition of the tropics, in estimates of the area of tropical forests, in the conversion factor from organic matter to carbon and conceptual differences. Brown and Lugo (1980) point out that, if their estimates are correct, predictions of the effects on the atmospheric CO_2 concentration of tropical deforestation would have to be revised, because the 1%/year estimate of Woodwell et al. (1978), which itself might not be accurate, would represent a significantly lower contribution of carbon to the atmosphere (50% less, according to Brown and Lugo). There are other implications of the lower carbon storage estimates, including revision of budget and model calculations.

Thus, some recent studies have suggested that man's activities may make tropical forests a net sink for atmospheric carbon. In addition, recent analyses of the oscillations in the Mauna Loa CO_2 record (Figure 3.1) indicate that the deciduous forests may be increasing in carbon content. Thus, there are very recent suggestions that the tropical and temperate forests could be storing carbon, at least over the past few decades. There is obviously much uncertainty about the biotic sources and sinks of carbon. It appears, however, that theoretical and observational studies of the ocean and biota indicate that the biota cannot at present be a net source of CO_2 as large as the fossil-fuel source. A recent survey made by the World Climate Programme (1981) adopted a value for net emissions from the terrestrial biosphere since early in the last century of between 75 and 175 Gt and an annual release at present of between 0 and 4 Gt.

(c) Future levels of atmospheric CO_2 concentration

There are considerable uncertainties about the role of the land biota in the carbon cycle and about the future rates of additions of anthropogenic CO_2. Nevertheless, there are strong indications that the rate of use of fossil fuels and the rate of interference with the terrestrial biota will not change rapidly and will continue to grow as the world population grows. In the absence of detailed knowledge of the carbon cycle, it could be assumed that at least during the next 20 to 30 years, the relative contributions of these processes to the increase of atmospheric CO_2 will probably remain essentially as they have been in recent decades. Thus Bolin (1979) suggests that, until the end of this century or so, the increase of atmospheric CO_2 will be fairly close to 50% of the fossil-fuel releases. Therefore, a continued annual increase of fossil-fuel combustion by 4% would give rise to an atmospheric CO_2 concentration of about 380 ppm (by volume) at the turn of the century, while a 2% increase in fossil fuel use would give a concentration of about 365 ppm (by volume).

Zimen (1979) suggests that, if current models of the carbon cycle are anywhere near correct, predictions can be made for a time horizon of 30 to 50 years. Assuming that the biota respond to the increased CO_2 in the atmosphere by absorbing more carbon but also that deforestation continues proportional to the global energy consumption in the future, Zimen calculates that we could continue to burn fossil fuels as long as the supply lasts without reaching a doubling of the preindustrial level of atmospheric CO_2, *if* the growth rate in fossil fuel use is less than 1% per year. With a 3% per year increase, Zimen calculates that a level of 150% of the preindustrial level would be reached in about 40 years and a doubling would occur by the year 2040.

Assuming that the biomass increases by up to 10% more than its initial carbon content, Niehaus and Williams (1979) have used a carbon model to illustrate the implications of various hypothetical energy strategies. They find that, as one extreme, when global commercial energy use is assumed to reach a maximum level of 30 TW in the next century (compared with about 8 TW today), and non-fossil fuel sources dominate the supply after the year 2000, then the atmospheric CO_2 would reach a level of 400 ppm (by volume) in the year 2050, by which time the emissions of fossil-fuel CO_2 are very small. In contrast, if the energy use is assumed to reach a maximum level of 50 TW and only fossil fuels are used, the atmospheric CO_2 concentration reaches almost 800 ppm (by volume) by the year 2050 at which time the annual fossil-fuel emissions are about seven times greater than those today.

Using the same model as that used by Niehaus and Williams (1979), Niehaus (in Energy Systems Program, 1981, p. 583) has examined the atmospheric CO_2 concentration implied by the two scenarios for energy supply developed by the IIASA Energy Systems Program (1981) and discussed very briefly in

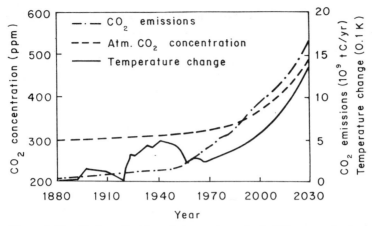

Figure 3.14 Carbon dioxide emissions, atmospheric CO_2 concentration, and computed temperature change for the IIASA high energy supply scenario. Source: Energy Systems Program (1981)

Section 3.4(a). Figures 3.14 and 3.15 show the CO_2 emissions and atmospheric CO_2 concentrations for the high and low scenarios, respectively. In addition, an estimate is given, based on climate model results of Augustsson and Ramanathan (1977), of the effect of the increase of CO_2 concentration on the global average surface temperature. The effects on climate are discussed in Section 3.5. Figures 3.14 and 3.15 show that the CO_2 emissions from fossil-fuel burning increase as a function of time in both scenarios, although they begin to level off in the low case. By the year 2030 the emissions are twice those of today in the low scenario and more than three times those of today in the high scenario. With these CO_2 inputs

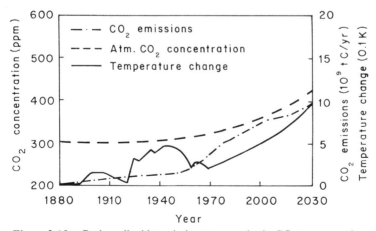

Figure 3.15 Carbon dioxide emissions, atmospheric CO_2 concentration, and computed temperature change for the IIASA low energy supply scenario. Source: Energy Systems Program (1981)

the carbon model calculates that the atmospheric CO_2 concentration would increase to about 430 ppm in the year 2030 in the low scenario and about 550 ppm in 2030 in the high scenario.

Since several energy projections have recently suggested that synthetic fuels could play an important role in future energy supply (e.g., Energy Systems Program, 1981), concern has been expressed that heavy reliance on synthetic (carbon-based) fuels will exacerbate the CO_2 problem. This concern was, in particular, expressed because the CO_2 release from synthetic fuels is larger per unit of energy released than that from conventional fossil fuels. For example, Rotty (1979) has calculated that the use of 1 million barrels per day of oil derived from coal would add 0.031 ppm CO_2 per year to the atmosphere; 1 million barrels per day of oil from shale would release 0.038 ppm CO_2 per year; and, in comparison, 1 million barrels per day of conventional oil would release 0.018 ppm CO_2 per year. The non-conventional oils release about twice as much CO_2 as the conventional oils per unit energy.

On the basis of these numbers, Rotty (1979) has investigated the CO_2 emissions implied by the second US National Energy Plan, produced in 1979. The relevant energy statistics from this plan are:

2.1%/yr growth in total energy to 1985
1.8%/yr growth in total energy 1985–2000
Non-conventional oil production of 1 million barrels/day by 1985,
Non-conventional oil production of 2 million barrels/day by 1990,
Non-conventional oil production of 3.5 million barrels/day by 1995,
Non-conventional oil production of 5 million barrels/day by 2000
US energy demand to grow at 2.1%/yr 2000–2020
US energy demand to grow at 2%/yr 2020–2050
After 2000, US non-conventional oil production continues to grow at 5%/yr.

Other assumptions had to be made regarding CO_2 releases elsewhere in the globe:

Global releases of CO_2 from fossil fuels continue to grow at the historic rate of 4.3%/yr until 2000 and 3%/yr thereafter. The release of CO_2 from deforestation will continue to be 2×10^{15} g carbon per year until 1990 and will decrease to 1×10^{15} g carbon per year during the period 1990–2000 and will be zero thereafter. The fraction of anthropogenic CO_2 that remains in the atmosphere is 0.4.

These assumptions were then used in a calculation of the atmospheric CO_2 concentration as a function of time during the period 1978–2050. Rotty (1979) has pointed out several consequences of these assumptions. Firstly, the amount of non-conventional oil produced in 2050 in this scenario is 60 million barrels per day. That is, the amount of non-conventional oil production is almost twice the present total energy use in the US. The assumptions led to an atmospheric CO_2 concentration of about 660 ppm (by volume) in the year 2050, a doubling of the present concentration. Less than 1% of the global anthropogenic CO_2 production is from the US non-conventional oil in 1990, and this proportion increases to 1.9% in 2025 and 3% in 2050. It has also been pointed out that the total energy use in this scenario is almost certainly higher than what will occur.

Overall the results suggested that even a massive non-conventional oil production in the US does not appear to enhance the global anthropogenic CO_2 production significantly. Therefore a second scenario was investigated, in which other nations also produced synthetic fuels, in proportion to their coal deposits. The rate of growth of synthetic fuel production was assumed to be the same as that given for the US. This resulted in 3.25 times as much synthetic oil. These assumptions led to a calculated atmospheric CO_2 concentration of about 676 ppm in the year 2050, and, as Rotty (1979) has pointed out, there is thus very little difference due to this large additional synthetic fuel production, especially compared with uncertainties regarding such factors as deforestation.

In a third scenario, Rotty (1979) assumed that by the year 2025 the world as a whole would be using 8 TW of solar, wind, biomass and hydroelectric energy and 4 TW of nuclear energy, which would be achieved by a 9% per year growth in non-fossil energy use. By 2045, therefore, fossil fuels would no longer be used. In this case the atmospheric CO_2 concentration grows at a rate similar to that in the first case until the year 2000, but after that date the CO_2 emissions become significantly smaller than those in the first scenario and the atmospheric CO_2 concentration reaches a steady-state value of about 500 ppm (by volume) in 2045.

The important message of these three calculations is, therefore, that it is apparently not important *which* fossil fuel is used for energy supply, because the mix cannot change quickly enough to make a significant difference in the resulting atmospheric CO_2 concentration, which is largely influenced by the choice between large-scale development of fossil fuels (conventional and non-conventional) or non-fossil energy suppliers (solar and nuclear). Such a conclusion is also evident, for example, in the study by Niehaus and Williams (1979). Likewise, it must be noted that the CO_2 concentration will also depend on the amount of energy growth foreseen. Rotty's scenarios, for instance, envisage a world energy use in 2025 of 32 TW, four times that of the present day. A halving of this would certainly have as large implications for CO_2 emissions as introducing a non-fossil energy supply.

The World Climate Programme (1981) concluded that the atmospheric CO_2 concentration in the year 2025 will be between 410 and 490 ppm with a most likely value of 450 ppm. The increase due to fossil-fuel emissions only would yield a concentration of 425 ± 25 ppm.

Keeling (1980) has discussed the future sinks for CO_2 and concludes that, if the amount of coal consumed during the fossil-fuel era is determined by physical limitations related to the energy required to extract and use the coal, then the amount of CO_2 produced would be eight or more times that of the preindustrial atmosphere. He concludes that this amount is so large that neither the oceans nor the biosphere, during the next several centuries, appear able to remove more than a small fraction of the added CO_2 from the atmosphere. Thus, Keeling suggests that the atmospheric CO_2 concentration will rise to a peak approaching that predicted if all of the CO_2 were to remain airborne. Only if growth in the consumption of fossil fuels is reduced to well below 1.5% per year is a substantially lower peak concentration likely, according to Keeling.

Hampicke (1980) has also computed the future atmospheric CO_2 increase in order to investigate how important the role of land ecosystems is. Eight scenarios were investigated. He concluded that any CO_2 quantity released from biotic sources in the future would be insignificant for future atmospheric CO_2 levels if fossil fuels are consumed in large amounts. On the other hand, if fossil-fuel use is constrained, CO_2 releases from the biota would be significant. Similarly, Hampicke concluded that biotic sinks such as reforestation would be significant only with low fossil-fuel use. He concluded further that, with high fossil-fuel use, the CO_2 increase would not be significantly reduced by lowering the airborne fraction. With low fossil fuel use, however, a lower airborne fraction would significantly reduce the atmospheric CO_2 increase if the net biotic inputs of CO_2 to the atmosphere would be reduced effectively within a short period of time.

Rotty and Marland (1980) have recently examined the possible constraints that could limit the use of fossil fuels and the subsequent production of CO_2. They suggest that there are three types of possible constraints: resource constraints; fuel demand constraints; and environmental constraints. They find that during the next 50 years there will be no real constraint as a result of a physical shortage of fossil-fuel resources. Further, they concluded that there will be no real constraint, at least during the early part of the 50-year period, as a result of the CO_2 problem, particularly without a more convincing case concerning the climatic changes caused by a CO_2 increase. Rotty and Marland conclude that the major constraint limiting the use of fossil fuels will be a slowing of the growth of fuel demand dictated by social/economic factors. They suggest that fuel-demand constraints will probably limit the use of fossil fuels to levels that keep the atmospheric CO_2 below 450 ppm for the next 50 years.

Because of uncertainties regarding the carbon cycle and the role of man's interference, especially with the land biota, it is not possible to make reliable predictions of the future atmospheric CO_2 concentration as a function of different energy supply scenarios. Nevertheless, by using assumptions regarding the carbon cycle, based on available observations, it is possible to make reasonable intercomparisons of different energy supply strategies in terms of the order of magnitude of their effect on the atmospheric CO_2 concentration. This section has described some of the many projections of future levels of atmospheric CO_2, based on simple analyses or models of the carbon cycle and energy supply. More projections are described, for instance, by Baes *et al.* (1976), NAS (1977a), Williams (1978), Elliot and Machta (1979), Bach *et al.* (1979), and Chan *et al.* (1980).

Most projections envisage a continued increase in the amount of atmospheric CO_2 due to the burning of fossil fuels and deforestation. Many projections suggest a doubling of the preindustrial atmospheric CO_2 concentration within the next 100 years.

3.5 THE EFFECT ON CLIMATE OF AN INCREASED ATMOSPHERIC CO_2 CONCENTRATION

(a) The 'greenhouse effect'

Many studies made with models of varying complexity have shown that increased atmospheric CO_2 produces a warming of the earth's surface and of the lower atmosphere. This warming is due to the fact that CO_2 is a good absorber/emitter of long-wave radiation. The surface warming is therefore caused by the increased downward emission of long-wave radiation from the CO_2 in the lower atmosphere. This effect has often been referred to as the 'greenhouse effect', although the analogy is not perfect. The warming of the lower atmosphere is due to the increased absorption of long-wave radiation from the surface and from clouds and to the increased downward emission from the CO_2 in the upper atmosphere.

(b) Feedbacks

Calculations show that a doubling of the atmospheric CO_2 concentration would lead to a net heating of the lower atmosphere, oceans, and land by a global average of about 4 W m^{-2} (Ramanathan et al., 1979; NAS, 1979). There is relatively high confidence that this net heating value has been estimated correctly to within $\pm 25\%$ (NAS, 1979). Greater uncertainties arise, however, in estimates of the change of global mean surface temperature ΔT resulting from the change in the heating rate ΔQ, because a number of feedback processes can increase or decrease ΔQ. This fact can be written in equation form as

$$\Delta T = \Delta Q / \lambda$$

where λ is a feedback parameter expressed in units of watts per square metre and per kelvin (W m^{-2} K^{-1}). NAS (1979) has discussed the feedback processes that influence ΔQ. In the simplest case, the earth is assumed to be effectively a black body and a doubled CO_2 concentration produces a temperature increase of 1 K (i.e., $\lambda = 4$ W m^{-2} K^{-1}). The most important feedback arises from the fact that the higher surface temperature leads to increased water vapour in the atmosphere. The increase of absolute humidity increases the absorption of long-wave radiation, thus adding to the effect of the CO_2 (positive feedback). There is also increased absorption of solar radiation by the water vapour. The result is that for doubled CO_2 the temperature increase would be 2 K (i.e., $\lambda = 2$ W m^{-2} K^{-1}).

One-dimensional radiative–convective models (see Chapter 2 for a description) give a value of $\lambda = 2.0$ W m^{-2} K^{-1} (Ramanathan and Coakley, 1978). However NAS (1979) points out that this value is uncertain by at least ± 0.5 W m^{-2} K^{-1}, because of uncertainties in possible changes in relative humidity, temperature lapse rate, and cloud cover and cloud height.

Another feedback process, shown in Figure 2.2, is the snow and ice albedo feedback. Estimates of this effect lead to a decrease of λ by 0.1–0.9 W m^{-2} K^{-1} (Lian and Cess, 1977; NAS, 1979).

Taking the above feedbacks into account, NAS (1979) estimated that λ is 1.7 ± 0.8 W m^{-2} K^{-1} and therefore that ΔT for doubled CO_2 is between 1.7 and 4.5 K, with 2.4 K a likely value. A review of more model results led to a conclusion similar to the one given in Section 3.5(e).

(c) Latitudinal and seasonal variations

Ramanathan *et al.* (1979) have pointed out that the radiative heating processes due to increased CO_2 are extremely dependent upon the atmospheric temperature and humidity distributions, both of which are seasonally and latitudinally dependent. The latitudinal and seasonal variations of the effect of increased atmospheric CO_2 were investigated by Ramanathan *et al.* (1979) using a model of radiative transfer and an energy balance (climate) model. The results from the radiative transfer model showed that, for a doubling of the CO_2 content, in winter the heating of the surface of the lower atmosphere (ΔQ above) decreased from 4.6 W m^{-2} at the equator to 2.2 W m^{-2} at 80°N, indicating the large latitudinal differences. Moreover, it was found that, northwards of 20°N, the heating is larger in the summer than in the winter and the difference between the summer and winter heating increases with increasing latitude. These effects are illustrated in Figure 3.16, which also shows that at high latitudes the Northern Hemisphere heating rates are larger than those of the Southern Hemisphere. Ramanathan *et al.* (1979) state that all of these differences in heating rates depend on temperature differences.

The radiative heating due to doubled CO_2 in winter is shown in Figure 3.17 separately for the surface and troposphere. The other seasons were found by

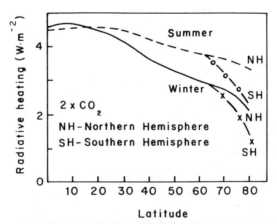

Figure 3.16 Radiative heating of the surface troposphere system due to a doubling of the CO_2 content. Assumed present-day value of the CO_2 mixing ratio is 320 ppm (by volume). Source: Ramanathan *et al.* (1979)

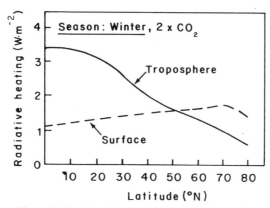

Figure 3.17 Breakdown of the total surface–troposphere heating due to doubled CO_2 into surface and troposphere heating. Source: Ramanathan *et al.* (1979)

Ramanathan *et al.* to have similar distributions. The reason for the difference between the tropics and polar latitudes is due to the influence of water, which is abundant in the tropical atmosphere and decreasingly abundant polewards. On the basis of this diagram, Ramanathan *et al.* speculate on the feedbacks that lead to the surface warming. They suggest that the surface warming in the tropics is produced by atmospheric feedback processes, while the increased CO_2 increases the atmospheric temperature. On the other hand, in high latitudes, the increased CO_2 mainly warms the surface by downward radiation.

In addition to the above effects on radiative heating, Ramanathan *et al.* found that the heating due to increased CO_2 is significantly different for clear sky and overcast sky conditions.

The calculations of heating rates were incorporated into an energy-balance climate model (see Chapter 2 for a brief description) to find the corresponding changes in surface temperature for the latitude zones and seasons. For a doubling of the atmospheric CO_2, the hemisphere average annual surface temperature increase was found to be 3.3 K. The annual surface temperature increased by just over 3 K in low latitudes to more than 5.5 K in high latitudes. Figure 3.18 shows that in the low latitudes there is virtually no seasonal variability in the temperature increase but in high latitudes the temperature increase is greater in spring at 65°N and spring and summer at 85°N. Ramanathan *et al.* find that the spring/summer amplification is entirely due to ice albedo feedback. As they point out, ice–albedo feedback requires the presence of solar radiation in addition to ice/snow. Although the authors recognize that the model does not include numerous time-dependent processes, the results nevertheless suggest that the spring/summer amplification could occur.

One important consequence of Figure 3.18 concerns the ability to detect in the future a climatic warming due to increased CO_2. Figure 3.18 suggests that the

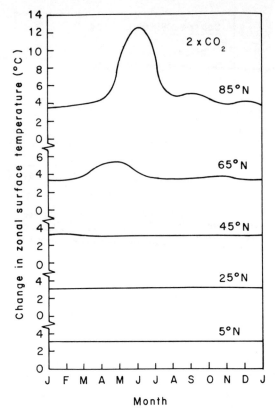

Figure 3.18 Increase in zonal surface temperature due to doubled CO_2 as a function of month for several latitudes. Source: Ramanathan *et al.* (1979)

change would be best observed in high latitudes in spring/summer. Indeed, Ramanathan *et al.* calculated that an increase of the atmospheric CO_2 concentration by a factor of 1.33 gave a Northern Hemispheric mean surface temperature increase of 1.45 K, but for June at 85 °N the increase was computed to be 6.5 K.

(d) Clouds

Model results referred to in the last section suggest that the heating rates due to increased CO_2 are influenced by the occurrence of overcast or clear sky conditions. In addition, clouds have other climatic effects which must be considered within the assessment of CO_2 effects. Clouds have two opposing effects on radiation (NAS, 1979; Ohring and Clapp, 1980). As cloud amount increases, the reflection of solar radiation increases, which would alone decrease the earth's surface temperature. At the same time, decreased upward long-wave radiation from the tropopause and downward radiation from the base of the clouds acts to raise the temperature of the earth's surface.

NAS (1979) concludes that the *net* result of increased low cloudiness, and very likely also middle cloudiness, is a cooling. The net effect of an increased amount of high cirrus clouds is less certain, but present estimates suggest a warming. Therefore, if the increased global temperature leads to an increased amount of low and middle clouds, there would be a negative feedback, whereas an increase of high clouds only would have a positive feedback. NAS (1979) concludes that, if clouds at all levels were increased by 1%, the net effect would be a cooling of about 0.4 W m^{-2}.

However, simple considerations such as a 1% increase at all cloud levels, are hardly likely to occur in reality. The cloud distribution is widely variable in space and time as a result of many other interacting climate factors. It is not known how the cloud distribution would change if the earth's surface and lower atmosphere warmed due to increased CO_2. In order to make such predictions, detailed numerical models are required; in addition, observations of cloud distributions in space and time are needed to validate the models.

Wetherald and Manabe (1980) have carried out a series of model simulations with the assumptions of either fixed or variable cloud cover to examine the sensitivity to increases in the amount of incoming solar radiation ('solar constant'). In fact, the response of the model with computed cloudiness to a 2% (or 4%) increase of the solar constant resembles the corresponding response to doubling (or quadrupling) the CO_2 content. The results of the solar constant experiments suggested that cloud feedback mechanisms have a relatively small effect on the sensitivity of area-mean and zonal-mean model climate. The total amount decreased in most of the region equatorwards of 50° latitude, with the exception of a narrow subtropical belt. However, it increased in the region polewards of 50°. Equatorwards of 50° latitude, the reductions of cloud amount and effective cloud-top height contribute to the increase in the effective emission temperature of the outgoing long-wave radiation and therefore enhance the cooling of the model atmosphere. At the same time, the reduction of cloud amount results in an increase of net incoming radiation, which contributes to the warming of the earth–atmosphere system. Polewards of 50° latitude, the increase of total cloud amount leads to a reduction of both incoming solar radiation and outgoing radiation. The changes of the two fluxes approximately compensated one another because of small insolation in high latitudes. The compensations mean that the cloud feedback mechanism had a relatively small effect on the sensitivity of the area-mean climate. However, Wetherald and Manabe (1980) point out that the method of cloud computation in the model is extremely simplified. Also the model itself did not consider the realistic geography of the global surface.

Paltridge (1980) has recently discussed the suggestion of NAS (1979) and Wetherald and Manabe (1980) that the cloud radiation feedback might reduce to negligible proportions of the increase in global surface temperature with increasing atmospheric CO_2. Paltridge (1980) points out that this statement is legitimate in view of the caveats given in the relevant studies, but might not be true, especially since it is based on the results of a limited number of experiments with numerical models, in which 'clouds' are generated by the model itself. Paltridge discusses what the results might be from a numerical model if (1) feedback via cloud optical depth is specifically included, (2) a different parameterization is used for generating total cloud amount, or (3) a different parameterization is used for generating boundary layer clouds. It is suggested that effects such as (1) and (3) may considerably reduce the quoted sensitivities of the global system to increasing CO_2. It would therefore appear necessary to include such effects in future modelling efforts. The results, however, support the contention of NAS (1979) that cloud feedback is unlikely to reverse the sign of the system sensitivity.

(e) General circulation model studies

The previous sections have shown that the response of the climate system to an increased CO_2 concentration cannot be expected to be uniform in space and time. Clearly, for a proper evaluation of the effects of increased CO_2, it is necessary to know not only what the change in global average temperature would be but also what changes in a number of meteorological variables (rainfall, temperature, wind direction, etc.) would occur in particular regions. One major effort to answer such questions is being made with the use of general circulation models (GCMs) of the atmosphere or atmosphere–ocean system. A brief description of such models is given in Chapter 2.

The first study made with a GCM was that of Manabe and Wetherald (1975). The numerical model was run with an atmospheric CO_2 concentration of 300 ppm in the first simulation and 600 ppm in the second simulation. Figure 3.19 shows the differences in latitudinally averaged temperatures between the two simulations. The global average surface temperature increase was 2.9 K, which was in line with estimates from simpler models of a 1.5–3 K increase for a doubling of the CO_2 concentration (Schneider, 1975). The temperature increase was considerably amplified in the polar region, where it was as much as 10 K. A further model result was that the global precipitation increased by 7%.

There were, however, several shortcomings as far as the model was concerned (discussed, for example, by Smagorinsky, 1977, and by the authors themselves). In particular, the distribution of oceans and continents was idealized and there was no heat storage by continents or oceans and no transport of heat by the ocean. The ocean was basically treated as a 'swamp' from which water could evaporate freely. In addition, the cloud distribution was fixed at present-day observed values in both simulations.

Manabe and Stouffer (1980) report using a global climate model to study the effect of quadrupling the CO_2 concentration in the atmosphere. The model consisted of a general circulation model of the atmosphere, a heat and water balance model of the continents, and a simple model of the mixed layer of the oceans. It had a global computational domain and realistic geography. Quadrupling the CO_2 content led to a general warming of the model troposphere and a cooling of the stratosphere. The warming was particularly pronounced in the lowest layers in

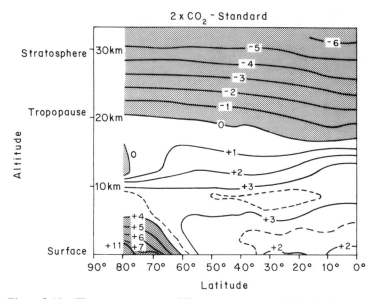

Figure 3.19 The temperature differences, averaged longitudinally, between a GCM simulation with current values of atmospheric CO_2 and a simulation with twice that amount. Source: Manabe and Wetherald (1975)

high latitudes and was relatively small in the tropics. The high-latitude warming of the Northern Hemisphere was significantly larger than that of the Southern Hemisphere. This interhemispheric difference in warming is accounted for by the fact that the contribution of the snow–albedo feedback mechanism is relatively small over the Antarctic continent. Manabe and Stouffer add that it is possible that the interhemispheric difference is exaggerated by the model owing to deficiencies in the sea-ice model.

Figure 3.20 shows the seasonal variation of the difference of zonal mean surface temperature between the $4 \times CO_2$ and $1 \times CO_2$ cases. In low latitudes the warming resulting from quadrupling the CO_2 content is relatively small and depends little upon season. In high latitudes the warming is larger and varies markedly with season, especially in the Northern Hemisphere. The results of Manabe and Stouffer show a maximum warming in early winter and small warming in summer. They point out that the polewards retreat of snow cover and sea ice is mainly responsible for the large annual mean warming in high latitudes, but the change of the thermal insulation effect of sea ice strongly influences the seasonal variation of the warming over the polar regions.

The increase of the global mean surface air temperature in the Manabe and Stouffer work was 4 °C, and they suggest that the corresponding warming due to a doubling of the CO_2 concentration would be approximately 2 °C. It is suggested that the smaller sensitivity of their model in comparison with the annual mean model of Manabe and Wetherald (1975) is partly due to the absence of or the

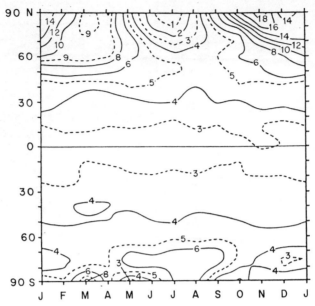

Figure 3.20 Latitude–time distribution of zonal mean difference in surface air temperature (kelvin) between the $4 \times CO_2$ and $1 \times CO_2$ experiments. Source: Manabe and Stouffer (1980)

smallness of the contribution from the albedo feedback mechanism over the Antarctic continent throughout the year and over the Northern Hemisphere in summer. The general warming of the atmosphere resulted in the enrichment of the atmospheric moisture content and an increase in polewards moisture transport. In the high latitudes the precipitation and run-off rates increased throughout the year.

Table 3.2 summarizes four GCMs that have been used recently to look at CO_2 effects; they were reviewed by NAS (1979). M1 refers to the model (and results) of Manabe and Wetherald (1975), described above. M2 is the model used by Manabe and Wetherald (1980). The primary difference from M1 is the inclusion of cloud feedbacks and a change of the snow and ice albedo criterion. M3 is reported by Manabe and Stouffer (1979, 1980). In this model, the geography is realistic, seasonal change is simulated and the ocean is treated as a 'mixed layer', i.e., the ocean has a heat capacity corresponding to that of an ocean mixed layer of constant depth but ocean transport is still not considered. Models H1 and H2 were developed completely independently from M1–M3. The main differences between H1 and H2 are in the treatment of the oceans. NAS (1979) points out that the model resolution of the H models is coarse and perhaps marginal for climate simulations, but these models take more physical processes into account.

The hemisphere mean surface temperature change was about 3 K in M1 and M2, the global mean was about 2 K in M3, 3.5 K in H1, and 3.9 K in H2. The

Table 3.2 Characteristics of five general circulation models used to study CO_2 impacts, λ is longitude, ϕ is latitude, T is temperature. Source: NAS (1979)

Model characteristics	Model predictions				
	M1[a]	M2[a]	M3[a]	H1[b]	H2[b]
Domain	$0° < \lambda < 120°$[c] $0° < \phi < 81.7°$	$0° < \lambda < 120°$[c] $0° < \phi < 90°$	Global	Global	Global
Land–ocean distribution	Ocean for $60° < \lambda < 120°$ $0° < \phi < 66.5°$	Ocean for $60° < \lambda < 120°$ $0° < \phi < 90°$	Realistic	Realistic	Realistic
Ocean	Swamp	Swamp	Mixed layer	Mixed layer	Swamp
Seasonal change	No	No	Yes	Yes	No
Cloud feedbacks	No	Yes	No	Yes	Yes
Snow and ice albedo	When $T < -25 °C$ 0.7 When $T > -25 °C$ 0.45 for snow 0.35 for ice	When $T < -10 °C$ 0.7 When $T > -10 °C$ 0.45 for snow 0.35 for ice	Depends on depth and underlying surface albedo For deep snow, 0.8 For thick ice, 0.7	For snow, depends on snow age, snow depth, underlying surface albedo, etc. For ice, 0.45	Same as H1
Horizontal resolution	About 500 km on a Mercator projection	$5°$ in longitude $4.5°$ in latitude	Spectral model with the maximum zonal wave number 15	$10°$ in longitude $8°$ in latitude	Same as H1
Vertical resolution	9 layers	9 layers	9 layers	7 layers	7 layers

[a] Models developed by S. Manabe and colleagues at the NOAA Geophysical Fluid Dynamics Laboratory, Princeton, NJ.
[b] Models developed by J. Hansen and colleagues at the NASA Goddard Institute for Space Studies, New York, NY.
[c] Cyclic continuity assumed at boundaries.

1 K difference within the M series has been attributed to the absence of seasonal changes and Southern Hemisphere effects in M1 and M2 (NAS, 1979). The differences between the M series and H series may be attributed at least in part to differences in the areas covered by snow and ice.

All of the models predicted a larger surface temperature increase at high latitudes, with a maximum change of 4–8 K in polar or subpolar mean annual values. Of the three models which calculate cloud distributions, M2 gave a decrease of high clouds in low latitudes, while H1 and H2 gave an increase. This could be due to differences in the treatment of cumulus convection (NAS, 1979).

NAS (1979) reviewed the model results in more detail and concluded that, on the basis of these results, a doubling of the CO_2 concentration would lead to a global mean surface temperature about 3 K, with a probable error of ± 1.5 K. Moreover, since the ice albedo feedback is greater in the Northern than in the Southern Hemisphere (because of the larger land area in the Northern Hemisphere and the lack of albedo change in the Antarctic), the temperature change is expected to be greater in the Northern Hemisphere.

It is pointed out by NAS (1979) that the GCMs produce time-averaged mean values of meteorological variables, such as temperature and rainfall, and these averages correspond reasonably well to those observed when global or latitudinal averages are considered. However, the regional distributions are not so well simulated, due to a considerable number of model shortcomings, including poor treatments of cloud, precipitation, and orographic effects. This led NAS (1979) to the conclusion:

> ... It is for this reason that we do not consider the existing models to be at all reliable in their predictions of regional climatic changes due to changes in CO_2 concentration.
>
> We conclude that the predictions of CO_2-induced climate changes made with the various models examined are basically consistent and mutually supporting. The differences in model results are relatively small and may be accounted for by differences in model characteristics and simplifying assumptions. Of course, we can only say that we have not been able to find such effects. *If* the CO_2 concentration of the atmosphere is indeed doubled and remains so long enough for the atmosphere and the intermediate layers of the ocean to attain approximate thermal equilibrium, our best estimate is that changes in global temperature of the order of 3 °C will occur and that these will be accompanied by significant changes in regional climatic patterns.

MacCracken (1980) and Washington and Ramanathan (1980) have compiled a number of the recent studies of the surface temperature change due to an increased atmospheric CO_2 concentration, including the studies listed above. Their compilation shows that one-dimensional radiative–convective models (e.g., Manabe and Wetherald, 1967), two-dimensional statistical–dynamical models (e.g., Potter, 1979), three-dimensional GCMs (e.g., Manabe and Wetherald, 1975), and GCMs with ocean–atmosphere interaction (e.g., GISS, 1978) produce similar ranges of temperature change.

The study by Gates and Cook (1980) has results that are somewhat different from those of the other studies. Gates and Cook used a GCM in which the boundary conditions varied seasonally and the ocean surface temperatures were kept equal to those of the present day. This assumes infinite ocean heat capacity with no feedback. The model's response is thus confined to the direct radiative effects of the increased CO_2, a situation that might arise during the next 20 years as the oceans are slow to respond. The result of a doubling (quadrupling) of the CO_2 was to give a global surface temperature increase of 0.3 K (0.5 K) in January. The warming is smaller than in other experiments, since the ocean surface temperature is not permitted to increase.

In July, the globally averaged tropospheric temperature increased by 0.33 °C (0.60 °C) for doubled (quadrupled) CO_2, with an average warming over land surfaces of 0.71 °C (1.04 °C). An interesting result of the simulations made by Gates and Cook is that there was no significant change in precipitation, and consequently little change in the atmospheric heating due to the release of latent heat. This is in contrast to the above-mentioned simulations with a 'swamp' ocean, in which both evaporation and precipitation increased in response to changes of sea surface temperature. Thus, Gates and Cook find that the model's simulated warming is less than that found in models with a swamp ocean, although the warming appears to be significant over the continents during the summer season.

Two recent studies have suggested that the increase of global surface temperature would not be of the order of 2 K or more for a doubling of atmospheric CO_2 concentration, as has been predicted by several GCMs. Newell and Dopplick (1979) and Idso (1980), using empirical approaches involving surface-energy-balance equations, have both suggested that the warming would be negligible. These studies both computed the direct surface radiative heating due to CO_2 increase and obtained the corresponding surface warming by multiplying by a 'response function' obtained from empirical formulae or observations.

Ramanathan (1981) has provided an explanation for the large difference between the results of the modelling and empirical approaches. He points out that the key to the problem lies in the proper understanding of the role of ocean–atmosphere interactions and feedback processes that produce a surface warming in climate models. He uses a one-dimensional model including both atmosphere and ocean that is able to reproduce the results of both the GCMs and the empirical approaches.

First, Ramanathan summarizes the effects on radiative heating of doubling the atmospheric CO_2 concentration. There are three basic effects:

- The increased CO_2 causes a radiative heating within the troposphere. In general, increasing the concentration of an optically thick gas such as CO_2 would cause radiative cooling in the low and middle troposphere. However, because of the overlap of CO_2 and H_2O bands in the 12–18 μm region, part of the enhanced CO_2 emission is absorbed by H_2O and leads to the net tropospheric heating.

● The downward solar flux at the surface decreases because of increased CO_2 absorption within the atmosphere. However, this decrease is much smaller than the increase in downward infrared radiation at the surface.

● The troposphere as a whole is subjected to a net global average radiative heating of ~3 W m^{-2} for a doubling of CO_2. This is about a factor of 3 larger than the surface heating. According to Ramanathan, Newell and Dopplick (1979) and Idso (1980) did not consider this tropospheric radiative heating.

The feedback mechanisms and processes that contribute to the surface warming have been shown schematically by Ramanathan (1981) as in Figure 3.21. There are three processes involved:

1. CO_2 direct surface heating is the change in the net downward radiative flux at the surface. This surface heating warms the ocean surface. The change in the mixed layer temperature is denoted by ΔT_m.
2. In this case, direct radiative heating warms up the troposphere, which

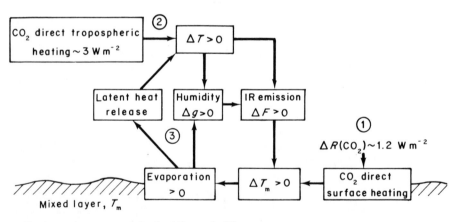

Numbers correspond to doubling of CO_2

	Process (1)	Process (2)	Process (3)	Total
Flux (W m^{-2})	1.2	2.3	12.0	15.5
Per cent	8.0	15.0	77.0	
ΔT_s (model dependent)	0.17	0.33	1.7	2.2

Figure 3.21 Schematic illustration of the ocean–atmosphere feedback processes by which CO_2 increase warms the surface. The contribution by the various processes to surface radiative heating, the percentage contribution and to the computed surface warming are shown in the table. All of the numbers correspond to hemispherically averaged conditions and apply to a doubling of CO_2. Source: Ramanathan (1981)

enhances the infrared radiation emission by the radiatively active constituents of the troposphere (H_2O, CO_2, O_3, clouds, other trace gases). The downward component of this enhanced infrared radiation emission amplifies the surface warming by process (1).

3. This process involves the interactions between ocean surface temperature, the hydrological cycle, and the convective adjustment process in the troposphere. The surface warming due to processes (1) and (2) enhances H_2O evaporation into the troposphere, which indirectly amplifes the surface warming in two ways: (a) The latent heat released within the troposphere (from the enhanced evaporation) warms the troposphere, thus enhancing the tropospheric infrared radiation emission. (b) The enhancement of evaporation also increases the absolute humidity of the troposphere, which increases tropospheric infrared radiation emission. Part of the increase in infrared radiation is emitted upwards and the remainder is emitted downwards to the surface. The latter downwards fraction amplifies the surface warming by processes (1) and (2). This feedback between temperature, H_2O, and infrared emission is largely controlled by ocean–atmosphere interactions, since the oceans are the primary source of atmospheric H_2O.

The table at the bottom of Figure 3.21 gives the magnitude of the processes as inferred by Ramanathan from model studies. Ramanathan adds that the snow/ice albedo feedback was not included in the analysis because it has a relatively small contribution, when compared with the processes in Figure 3.21, to globally averaged surface warming.

Ramanathan concludes that this ocean–atmosphere feedback is included implicitly in most model studies, although model results have not usually been analysed in terms of this interaction. Consequently, several empirical approaches based on surface-energy-balance considerations, unaware of the feedbacks that contribute to the surface warming, have obtained results that contradict the model results.

(f) The role of the oceans

The models that have been used so far to investigate the effects of increased CO_2 have not treated the interaction of the oceans with the atmosphere specifically, or they have treated the interaction in a very simplified manner. In particular, in neither the 'swamp' model nor the 'mixed layer' treatment is there any horizontal transport of heat by the oceans. Observations of the present-day oceans, however, suggest that at some latitudes the oceans transport as much as 50% of the polewards heat flux of the atmosphere–ocean system (Oort and Vonder Haar, 1976). Thus, it would seem that omission of this transport in climate models could mean that the results of CO_2 sensitivity tests are not realistic.

As NAS (1979) has pointed out, the atmospheric models suggest that the warming at high latitudes will be larger than at low latitudes. If this leads, as theory suggests, to a change in the atmospheric state such that wind stress at the ocean surface is reduced, then it is possible that the ocean polewards heat flux would be reduced. Because there is a required total heat balance of the whole system, the atmosphere would then be required to compensate for the reduced ocean transport and this would lead to a reduction of the equator-to-pole temperature gradient and a reduction of the larger temperature increase in the polar area.

A recent workshop analysed in more detail the sensitivity of the ocean to changes in the atmosphere due to a CO_2 increase (AAAS, 1980, pp. 1–23). It was concluded that a general slackening in wind stress by 40%, coupled with a warming of several degrees centigrade and possibly reduced salinity of the surface layers of the ocean in high latitudes, would lead to an overall slowdown in the circulation of water in the ocean and a tendency for the surfaces of constant density to be more nearly horizontal than at present. The working group suggested that over the long term the surface waters that are cooled at high latitudes would tend to sink to intermediate depths rather than to form deep water, and the compensatory near-surface return flows from other latitudes would be modified.

When the ocean has been taken into account in model studies of CO_2 effects, only the mixed layer (the top 100 m or so) has been considered. The mixed layer responds on an annual time scale, storing and releasing heat as a function of season. On a longer time scale, say of decades, the interactions and heat exchanges with deeper layers of the oceans must be considered. NAS (1979) points out that, over longer time scales, the effective thermal capacity of the ocean for absorbing heat is nearly an order of magnitude greater than that of the mixed layer alone. Thus, with regard to the CO_2 effects, if the oceans do have a larger capacity than just that of the mixed layer, then the equilibrium global surface temperature change in response to, say, a doubling of the CO_2, will be reached after the doubling has occurred; NAS (1979) suggests that the delay could be of the order of a few decades. The implications of this delay are discussed further in the next section.

Washington and Ramanathan (1980) have discussed further atmosphere–ocean interactions of significance to the CO_2 effects. One point they make is that, if a warmer and wetter atmosphere is the result of increased CO_2, then it could be expected that warmer and less saline (less dense) water would form at the surface. This would result in less formation of cold intermediate and bottom water. Washington and Ramanathan (1980) suggest that for simulation of CO_2 effects it is likely that the *entire ocean needs to be taken into account*, not just the surface layer.

Thus, it appears that coupled atmosphere–ocean general circulation models will be required for more realistic studies of the sensitivity of the climate system to an increased atmospheric CO_2 concentration. It has been pointed out, for example by Washington and Ramanathan (1980), that coupling the atmosphere and the ocean in a model is difficult because of the vastly different time scales. In the so-called 'swamp' and 'mixed layer' models, the atmosphere and ocean components are run simultaneously and the information required by either is communicated between them. This cannot be done when levels much deeper than the ocean mixed layer are considered, because of the time taken for the models to come into equilibrium. For example, the swamp model of Manabe and Wetherald (1975) took about 300 days of simulated time to come into equilibrium and the mixed layer model of Manabe and Stouffer (1979) took about 10 years of simulated time. If the complete ocean were simulated, thousands of years of simulated time would be required to bring the models into equilibrium. Computer time then becomes a powerful constraint. The present solution (e.g., Manabe et al., 1979; Washington et al., 1980) is to run the models asynchronously, but the methods are not without problems and it is not clear that a true equilibrium has been reached in which the results are the same as would be found with synchronous coupling (Washington and Ramanathan, 1980). Research is continuing on the problems of coupled ocean–atmosphere models, which are necessary for studies of the effects of an increased atmospheric CO_2 concentration.

(g) The transient response to a CO_2 increase

The climate modelling studies that have been discussed so far in this chapter have all considered the equilibrium climate response to a step-function change in the CO_2 concentration. For instance, the CO_2 has been doubled and the difference between two equilibrium states of the model (with and without the CO_2 doubling) has been investigated. In reality, the CO_2 concentration is increasing year by year and it is more realistic to study the *transient* response of the climate to the time-dependent CO_2 increase.

Recently, several studies have investigated the time lags in the climatic response to forcing. For instance, Hasselmann (1979) suggested that the effect of the heat capacity of the deeper layers of the ocean could delay by many decades the approach to equilibrium of the climate system when perturbed by CO_2. Thompson and Schneider (1979) estimated that the globally averaged surface temperature change lags 5–20 years behind that computed with a 'swamp' model. NAS (1979), referred to in Section 3.5(f), also noted that a delay of up to several decades could occur because of the increased heat capacity associated with the exchange of upper level waters with deeper oceanic reservoirs.

Estimates of the lag in response due to the thermal capacity of the oceans have been made on the basis of simplified models of the ocean–atmosphere system. Hunt and Wells (1979) estimated that the time lag in the atmospheric temperature change would be eight years. They made this estimate by using a model of the oceanic mixed layer down to 300 m. Hoffert et al. (1980) used a one-dimensional model of the ocean that included upwelling, diffusion, and an inflow of cold water at the base. This model suggested that deep-sea thermal storage could delay the temperature change predicted by atmospheric models by 10 to 20 years in the period 1980 to 2000. Schneider and Thompson (1981) estimated that the temperature lag would be 5 to 20 years, depending on the rate at which the upper layers of the ocean mix with water from the lower layers and on the sizes of these bodies of water. Hansen et al. (1981) have shown the effect of ocean thermal capacity in suppressing a CO_2 warming with the aid of the box-diffusion model of Oeschger *et al.* (1975). The calculated warming due to the CO_2 increase between 1880 (CO_2 amount assumed to be 293 ppm) and 1980 (335 ppm) was ~0.5 °C if ocean heat capacity was neglected. The inclusion of the heat capacity of the mixed layer reduced the warming to ~0.4 °C. The inclusion of the diffusion of heat into the thermocline or the deep ocean further reduced the CO_2-induced warming to ~0.2 °C. Hansen *et al.* emphasize, however, that the ocean heat capacity may delay the full impact of the CO_2 warming for a few decades but, since man-made increases in atmospheric CO_2 are expected to persist for hundreds of years, the warming will eventually show up.

Most studies suggest that the globally averaged surface temperature would reach its equilibrium value several (or many) years after the doubling of the CO_2 concentration.

Schneider and Thompson (1981) extended this work by considering the approach of latitudinally averaged temperature to the equilibrium value, using a hierarchy of simple energy-balance models. They found that the approach to equilibrium in the various latitudes differs from the global average approach. Schneider and Thompson (1980) suggest that the time evolution of regional climatic anomalies may well be different from that indicated by equilibrium climatic modelling experiments. For example, a number of projections suggest that there will be about a 20% increase in the CO_2 concentration by the year 2000. It may not be enough simply to take one-fifth of the equilibrium response of a model to a CO_2 doubling and expect this to be the climatic signal occurring around the year 2000. Firstly, one could expect a time lag in the response to the CO_2 increase, and secondly, the response is transient and evolves differently from what could be derived from results of equilibrium models. Schneider and Thompson (1981) conclude that better accounting of the time and space variations of thermal inertia is a high priority item for research on the effects of increased CO_2.

(h) The question of signal-to-noise ratios

As indicated in Chapter 2, for instance, the natural climate has an inherent variability. This is well illustrated by the fact that one winter is not a replica of the previous winter, and that the average temperature changes from decade to decade. As Mitchell (1979) has pointed out, this variability has two implications as far as the effect of a CO_2 increase is concerned. Firstly, the inherent variability of climate is likely to continue in the future at the same time that the CO_2 is being added, and would tend to obscure the climate change that is properly attributable to the CO_2 increase. Secondly, the many natural causes of climatic variability would in the future compete with the added CO_2 to produce future changes of climate.

Two recent papers have looked at the implications of this inherent variability of 'climatic noise' for the detection of climatic change due to increasing CO_2. Madden and Ramanathan (1980) made a quantitative estimate of the noise and examined the feasibility of detecting a CO_2 signal. The noise level was estimated on the basis of nearly continuous records of observed monthly mean temperatures from 1906 until 1977 for 12 stations close to latitude 60°N. Two possible responses of the climate system to a CO_2 increase were considered. In the first case it was assumed that the feedback parameter λ (see Section 3.5(b)) equals 4 W m⁻² K⁻¹. Madden and Ramanathan suggest that this is an upper limit for the value of λ, and this gives the lower limit for the signal due to a CO_2 increase. This lower limit is referred to as the zero feedback case. Results of the Manabe and Wetherald (1975) GCM experiments were taken to estimate the upper limit for the signal that is referred to as the positive feedback case. For the estimate of the future increase of CO_2 concentration, the projections of Baes et al. (1976) were adopted. Fast growth and slow growth scenarios were considered, in which the growth rates of fossil-fuel consumption were 4% and 2%, respectively. To account for the range of CO_2 signal due to uncertainty about future airborne levels, Madden and Ramanathan considered the fast growth case with 60% airborne as an upper limit and the slow growth case with 40% as a lower limit.

Figure 3.22 shows the results of the analysis. The noise level is that calculated for the summer season. The averaging time in years for the computation of the

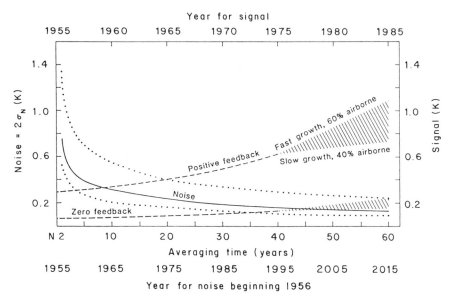

Figure 3.22 Estimate of noise level in summer season (solid line) and 95% limits (dotted lines) plotted against averaging time and year beginning in 1956. Positive feedback signal (upper dashed line) and zero feedback signal (lower dashed line) are plotted against the year indicated at the top. Possible delays due to inertia of the climate system are not accounted for in the signal. Fanning out of the signal curves indicates the uncertainty of future CO_2 levels in the atmosphere; the upper values are based on fast growth with 60% CO_2 airborne and the lower values on slow growth with 40% airborne. Source: Madden and Ramanathan (1980)

noise level is shown below the figure, along with a date based on the assumption that averages begin in 1956. The signals are plotted against the time scale at the top of the figure and neglect any delays that might arise from the inertia of the climate system. Figure 3.22 indicates that if the positive feedback models predict the effects of CO_2 correctly, a nine year average of summer temperatures computed from 1956 until 1964 would come from a distribution that is 2 standard deviations higher than the temperature before there was any effect due to increasing CO_2. On the other hand, if the zero feedback models are correct, a 45 year average from 1956 until 2000 would be required for a similar situation to occur. Therefore, Madden and Ramanathan find that the range of time when we might be able to establish statistically that model predictions are correct extends from the present if positive feedback models are correct, to the year 2000 if zero feedback models are correct. The 20 year average temperature from 1956 until 1975 is, in fact, not higher than the 20 year average temperature from 1906 until 1925. So, there is no statistical evidence that there has been an effect due to increasing CO_2 on the mean zonal temperature at 60°N. Therefore, Madden and Ramanathan conclude that either the models overpredict or other compensatory climatic changes are occurring. If the latter is not a serious problem and the warming is occurring at a rate lower than that predicted by GCMs but higher than that predicted by zero feedback models, then the signal should be detectable between now and the year 2000. Madden and Ramanathan find that it will be easiest to detect effects in summer data. Finally, they point out that unambiguous statistical proofs regarding the effects of CO_2 based on observations of a single variable will always be difficult and the effects should be looked for in several variables simultaneously.

The second recent study on the question of signal-to-noise ratios was made by Wigley and Jones (1981), who estimated the CO_2 signal on the basis of GCM results of Manabe and Stouffer (1980), in which the signal is greatest at high latitudes and in late autumn to winter. Figure 3.23 shows the signal-to-noise ratio as a function of latitude and month as computed by Wigley and Jones (1981). The temperature changes corresponding to a doubling of CO_2 are highly significant for all months and latitudes relative to the noise level. Figure 3.23 shows that the effects of CO_2 should first be detectable in the summer months in middle latitudes, although the maximum signal is in winter in high latitudes, because of the lower noise level in summer in middle latitudes.

Wigley and Jones examined the hypothesis that CO_2 was a causal factor in the early 20th-century warming (see Figure 2.4). In order to do this, a comparison was made between the spatial and seasonal patterns of the observed warming and those expected on the basis of CO_2 modelling experiments. In fact, the comparison was sufficiently good that it could be taken to support CO_2 as a cause for the early 20th-century warming, although Wigley and Jones point out some interesting differences. This similarity between natural and predicted CO_2-induced temperature changes suggests that analysis of spatial and/or seasonal details of future variations of surface temperature might be of little value in distinguishing natural changes from the effects of CO_2. Wigley and Jones conclude by pointing

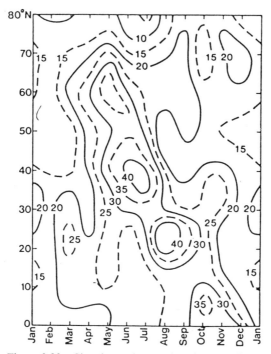

Figure 3.23 Signal-to-noise ratio for predicted CO_2-induced changes in surface air temperature as a function of latitude and month. The signal is based on the numerical modelling results of Manabe and Stouffer. The noise has been calculated from grid point surface temperature data: the value for month j at latitude L is the areally weighted average of grid points at $L - 5$, L, and $L + 5$, and the noise level is proportional to the standard deviation of month-j values over the period 1941–1980, corrected for autocorrelation effects. Source: Wigley and Jones (1981)

out that the effects of CO_2 may not be detectable until around the turn of the century. The atmospheric CO_2 concentration will probably have become sufficiently high by this time that, according to Wigley and Jones, a climatic change significantly larger than any that has occurred during the past century could be unavoidable. They therefore suggest that it is possible that decisions will have to be made (for example, to reduce anthropogenic CO_2 emissions) some time before unequivocal observational 'proof' of the effects of CO_2 on climate is available.

Hansen *et al.* (1981) have computed the observed surface air temperature trend for the globe for the period 1880–1980. They found that a global temperature was almost as warm in 1980 as it was in 1940 and suggested that the common conception that the world is cooling is based on Northern Hemisphere experience to 1970. The time history of the warming over the past century did not follow the course of the CO_2 increase, indicating that other factors substantially influence the global mean temperature. Figure 3.24 shows the comparison made by Hansen *et al.* (1981) between the observed global temperature trend and that computed by a one-dimensional radiative–convective atmospheric model coupled to an ocean. The effects of volcanoes and changes in solar luminosity have been included, in addition to the effects of CO_2. Hansen *et al.* concluded that the basic agreement between the observed temperature trend and that produced by the model suggest that CO_2 volcanic aerosol, and solar luminosity changes are responsible for much of the global temperature variation over the past century. They also found that the CO_2 warming should rise convincingly above the 'noise' level of natural climatic variability by the end of the century.

Kukla and Gavin (1981) have studied observations of the extent of polar ice during different seasons since the 1930s and found qualitative agreement with the expected effect of CO_2. (The work of these authors has also been reported in the *Bulletin of the American Meteorological Society* (1981), **62**, p. 1607.) The latter report states that Kukla and Gavin found that the extent of pack ice around Antarctica during summer has decreased in the 1970s compared with the 1930s.

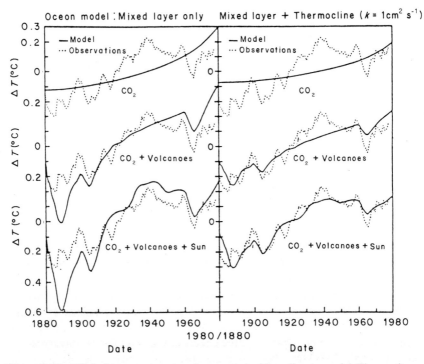

Figure 3.24   Global temperature trend computed with a climate model. The results on the left employ a 100 m mixed layer ocean for heat capacity, while those on the right include diffusion of heat into the thermocline to 1,000 m. The zero point of ΔT is defined such that the mean ΔT is zero for both observations and models. Source: Hansen *et al.* (1981)

They also found that in the Northern Hemisphere the 1974–1978 average surface air temperature in the zone of melting snow in spring and summer was up to 0.9 °C higher than during the 1934–1938 period. During the autumn and winter the temperatures in this zone were about 5 °C lower than in the 1930s. Kukla and Gavin suggest that studies of the highly variable components of the climate system such as snow and pack-ice fields and the associated surface temperatures should be given priority in the search for early signals of a CO_2 effect. They point out, however, that it is presently not known to what extent the observed changes in snow and ice areas and in surface temperatures can be explained by natural fluctuations or by processes unrelated to CO_2.

(i) Scenarios of past climate

Since there are many uncertainties in the results of model experiments to investigate the effect of increasing atmospheric CO_2 concentration and since it is of interest to 'predict' the regional climate changes and particularly the precipitation changes due to such a potential increase, other approaches to the problem have also been taken. Kellogg (1977) has suggested that one way to find out what a warmer earth might be like is to study climatic evidence from periods when the earth was warmer than it is now. As Kellogg points out, such a time existed about 4,000 to 8,000 years ago, a period he refers to as the 'Altithermal'. Evidence for climate conditions at that time can be derived from the distribution of fossil organisms in ocean sediments and pollens in lake sediments, the history of lake levels, extent of mountain glaciers, tree-ring widths, etc. (see, for example, Lamb, 1972, 1977). Figure 3.25 shows a reconstruction made by Kellogg of the conditions during the Altithermal, based upon geological and palaeobiological evidence. The map refers to precipitation relative to the present and is not, of course, complete. Kellogg points out, for example, that North Africa was generally more favourable for agriculture than it is now, that Europe was wetter, Scandinavia drier, and a belt of grassland extended across North America in what subsequently became forest land.

Kellogg rightly cautions the reader that this map must not be accepted as a literal representation of what might occur if the earth becomes warm again, since the causes of the warming could have been quite different from those of the potential warming due to anthropogenic changes. The causes of the Altithermal warmth are unclear. Another problem, not explicitly addressed by Kellogg, is that the Altithermal Period was time-transgressive. That is, the peak warming did not occur at the same time in all places. particularly in the Southern Hemisphere, where maximum warmth may have occurred between 9,000 and 7,000 years BP (Pittock and Salinger, 1981). Thus it is likely that the conditions plotted, for example, in Figure 3.25 did not all occur at the same time, even though they do individually represent the conditions of a warm period that occurred during the present interglacial period.

This time-transgressive nature has been pointed out, for example, by Flohn (1980). The maximum of the last glacial period occurred 22,000–18,000 years

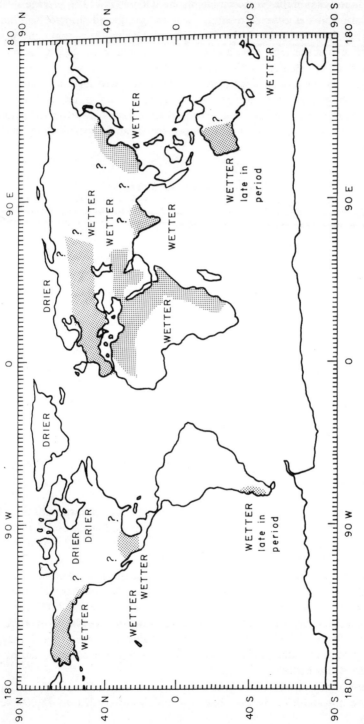

Figure 3.25   Schematic map of the rainfall anomalies, predominantly during the summer, during the Altithermal Period (4,000–8,000 years ago). The terms 'wetter' and 'drier' are relative to the present. Blank areas indicate lack of information. Source: Kellogg (1977)

ago. After that, the climate started to warm up, but not at the same rate everywhere. The Scandinavian ice disappeared about 8,000 years BP, the Labrador ice disappeared about 4,500 years BP and some ice, e.g., on Baffin Island and Greenland, still exists to the present day in the Northern Hemisphere. Flohn suggests that Eurasia and Africa experienced their warmest epoch of the last 75,000 years during the Altithermal but eastern North America remained relatively cool, certainly during the summer.

Flohn (1980) has extended the approach taken by Kellogg (1977). He suggests that the 'level of perception' of a warming is when the global surface temperature increase is 0.5 K. A warming of 1 K would be equivalent to the medieval warming that had its peak around 900–1050 AD. A warming of 1.5 K would be equivalent to the postglacial warm period, referred to above as the Altithermal Period, which Flohn dates as about 5,500 to 6,500 years before the present. A warming of 2–2.5 K would be equivalent to the last interglacial period, referred to as the Eem and dated about 125,000 years before the present. Lastly, Flohn considers a scenario of a 4 K warming, in which the Arctic Ocean is assumed to be ice-free. Flohn cites data suggesting that the Arctic Ocean has not been ice-free in the past 2.3 million years. Flohn (1980) made an analysis of the climatic conditions during each of these periods, but also asked the question: Can climatic history repeat itself?

This question is of major importance, because two boundary conditions that existed during the Altithermal Period have changed basically. Firstly, the presence of limited and shallow but not negligible permanent ice sheets in eastern Canada during the Altithermal gave an asymmetry to the circulation which would not occur today. Secondly, since the Altithermal, the climatic boundary conditions have been increasingly changed by desertification effects. Because of these changes, Flohn has concluded that, for example, along the northern margins of the African and Asian arid belt, no substantial increase of rainfall should be expected. At the southern flank some increase might be possible, if (as expected) the intensity of the subtropical anticyclones weakens together with their displacement towards higher latitudes. Flohn likewise points out that, because of man-made desertification effects, reconstruction of the natural vegetation cover, under present population pressure, might be delayed by several decades until a reliable long-term increase of rainfall could be achieved.

Two recent studies have used instrumental observations of temperature, rainfall, and pressure during the present century as a basis for discussing the response of the climate system to a warming (Williams, 1980; Wigley et al., 1980). Williams (1980) has looked at regional rainfall, temperature, and pressure anomalies in the Northern Hemisphere for summers and winters within the last 70 years when the Arctic was warm. The reason for choosing warm Arctic seasons arises from the model and observational studies that have indicated the Arctic to be more sensitive to climatic changes, although it should be noted that the Arctic temperatures also exhibit more inherent variability. Clearly, the warm Arctic seasons were not a result of a CO_2 increase; rather, they were the result of non-linear interactions between components of the climate system. The relevance of looking at warm Arctic seasons has been discussed in Chapter 2 (Section 2.7(d)).

84

Figure 3.26, from Williams (1980), shows the differences in surface temperature between the 10 warmest Arctic winters and the remaining winters in the 1900–1969 period. Shaded areas indicate regions where the 10 winters averaged together were cooler than the long-term mean. The warmth of the Arctic area is dominated by a 3 °C anomaly around 90°E. In one area of the Arctic the average temperature of the 10 winters is lower than the long-term mean. Elsewhere, in the belt around 50°N, temperatures are also higher than the long-term mean. Around the Mediterranean, from India to Japan, and from Mexico into the Caribbean, temperatures are lower than the long-term mean, though nowhere by more than 1 °C. It was found that the positive temperature anomalies in Scandinavia, Alaska, and the American midwest are statistically significant and the negative anomalies in the India area, Japan, and northeastern Canada are also statistically significant at a number of stations.

Figure 3.27, from Wigley *et al.* (1980), shows the differences in mean annual surface temperatures between the five years during which the Arctic was warmest and the five years when it was coldest. The largest warming occurs in the same

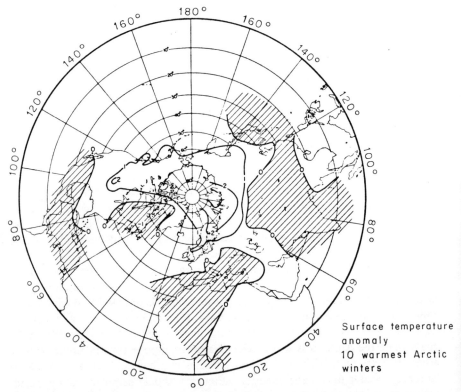

Surface temperature
anomaly
10 warmest Arctic
winters

Figure 3.26   Differences in surface temperature between the mean of the 10 warmest Arctic winters and the long-term mean of the remaining winters in the 1900–1969 period. Shaded areas indicate negative anomalies. Source: Williams (1980)

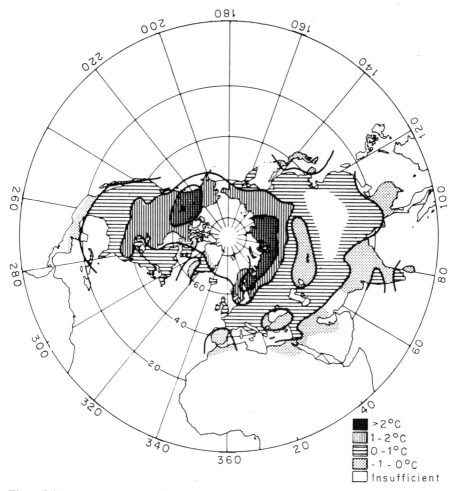

Figure 3.27   Mean annual surface temperature changes from cold to warm years. The corresponding change in the hemispheric mean temperature is 0.6 °C. Source: Wigley *et al.* (1980)

area found for the 10 warmest winters by Williams (1980). Likewise the band of warming in the American midwest is similar, and Wigley *et al.* also found a cooling in the Mediterranean area, India, and Japan.

One would not expect the distributions in Figures 3.26 and 3.27 to be exactly alike, partly because one is based on annual conditions, while the other is based on winter conditions and also because different years were considered and the anomalies are differently computed. Nevertheless, there is some similarity between the two maps, partly because, as both studies found, the anomalies in the winter are much larger than those in the summer, and the winter anomalies thus dominate the annual averages shown in Figure 3.27 from Wigley *et al.*

86

Figure 3.28 shows the differences in precipitation over North America between the 10 warmest Arctic summers and the long-term mean (Williams, 1980). The shading indicates where the precipitation decreased. However, since precipitation is influenced considerably by factors such as local geography, the absolute values of the precipitation differences have not been plotted. Alaska and the west coast of the United States are drier and there is an extensive area of negative anomalies in the central and eastern parts of the map. The southern states and Florida have increased precipitation. Whereas, in winter, only one station appeared to show a statistically significant difference between the mean precipitation of the 10 warmest seasons and the long-term mean, for the summer season statistically significant differences were found at 15 stations. The increased wetness in the southern states and Florida and the increased dryness in the midwestern region

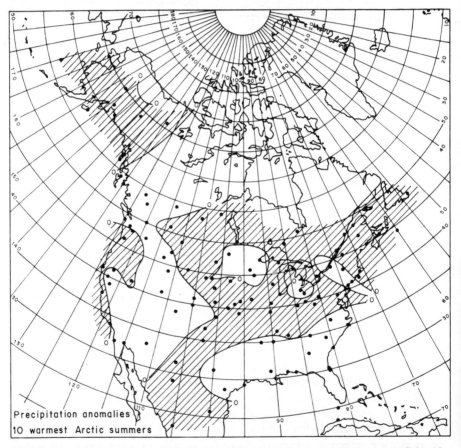

Figure 3.28   Differences in precipitation in North America between the mean of the 10 warmest Arctic summers and the long-term mean of the remaining summers in the 1900–1969 period. Shading indicates where precipitation decreased. Source: Williams (1980)

are statistically significant. It is interesting to note that the area of significant decrease of precipitation in the midwest roughly corresponds with the only area in North America where the summer temperature difference for the same sets of years is greater than +1 °C.

The annual precipitation differences between the five years with the warmest and the five years with the coldest Arctic temperatures are shown in Figure 3.29 taken from Wigley *et al.* (1980). The dryness in midwestern North America is common to both studies. Wigley *et al.* (1980) suggest that the most important precipitation differences are the decreases over much of the USA, most of Europe, Russia and Japan and the increases over India and the Middle East. Wigley *et al.* point out that, although annual precipitation in Europe and Russia is less in warm years, the potential effect of any such reductions in rainfall depends on their timing. They find that the largest reductions occur mainly in winter and autumn

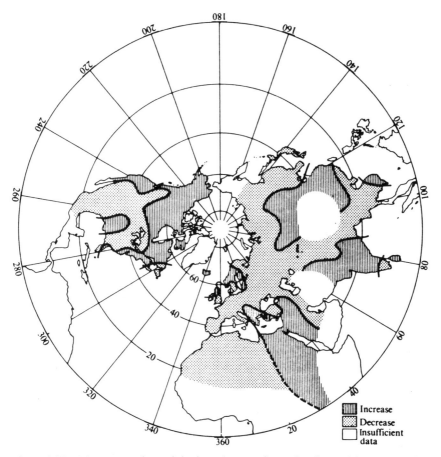

Figure 3.29   Mean annual precipitation changes from the five coldest to the five warmest years in the period 1925–1974. Source: Wigley *et al.* (1980)

but that most regions also show reductions in summer precipitation. A few stations (Lyons, Madrid, Lisbon, Vienna) experienced increased rainfall in warm years.

Neither the study of Williams (1980) nor that of Wigley *et al.* (1980) claims that the distributions of temperature and precipitation anomalies can be taken as predictions of the changes to be expected due to a doubling of the CO_2 concentration of the atmosphere. However, they should be useful in guiding the development of scenarios of potential changes. An advantage of this type of analysis is that it illustrates clearly that large coherent anomalies are a basic response to climatic forcing and that seasons respond differently owing to the different climatic processes that dominate in each season. It is likely that an improved understanding of the response of the climate system to perturbations, such as a doubling of the atmospheric CO_2 concentration, will come from a combination of such observational and model studies.

(j) Effects of a CO_2 increase on snow and ice cover

It has frequently been pointed out that a warming due to an increased atmospheric CO_2 concentration would have an effect on the snow and ice cover. Barry (1978) has stressed that the different forms of ice and snow would each respond differently to a warming. For instance, in low and middle latitudes warming would decrease the frequency of snowfall (more falling as rain) and the duration of snow cover on the ground, whereas in high latitudes, where snowfall is limited by the low vapour content of the air due to the low temperatures, a warming would give more snowfall, but the duration of snow cover is likely to be only marginally affected.

The World Climate Programme (1981) points out that an atmospheric warming would reduce the extent of frozen ground in polar latitudes. However, the time taken to melt deeper layers of permafrost extends over millennia. The same report states that the snow line of mountain glaciers rises about 100 m in elevation per centigrade degree of warming and that a general warming of 2 °C would melt most existing cirque glaciers, while a warming of 4 °C would cause smaller mountain glaciers to melt.

The effects of a warming on the ice sheets of Greenland and Antarctica are likely to be complex and on a much longer time scale than considered for snow and sea ice. Melting of the ice sheets would lead to a rise of world sea level, but generally this is considered to take place on a time scale of hundreds to thousands of years. There has been much discussion of the West Antarctic ice sheet in this regard. Hughes (1975) suggested that this ice sheet is unstable and Mercer (1978) sees a warming associated with atmospheric CO_2 as a possible cause of deglaciation in the future. Hollin (1970) postulates that major ice sheet 'surges' would lead to 10 to 30 m rises of sea level in less than 100 years. However, Barry (1978) has pointed out that the rate of propagation of a surface warming to the base of a several-thousand-metre-thick ice sheet is extremely slow, and any direct response of the ice to a climatic warming should be considerably lagged. The World

Climate Programme (1981) states that most glaciologists do not support the suggestion that CO_2-induced warming of the ice surface could initiate rapid disintegration of the West Antarctic ice sheet during the next century.

The question of the effect of a CO_2-induced warming on the Antarctic ice sheet has been discussed in more detail recently by the members of a workshop (AAAS, 1980, pp. 10–12). They emphasize that only a part of the ice sheet is vulnerable to any climatic warming expected from a doubling or quadrupling of the atmospheric CO_2 levels during the next 100 years or so. The vulnerable part is the one that is grounded far below sea level and can persist only so long as it is buttressed by fringing rock thresholds or ice shelves. This applies to most of the West Antarctic and part of the East Antarctic. The workshop participants suggested that the West Antarctic ice sheet at present seems to be well insulated against warming as the midsummer air temperature along its iceshelf front is about −5 °C. Moreover, they point out that ample warning of the southwards advance of a climate unfavourable for the West Antarctic would be available from satellite monitoring of the small ice shelves on the Antarctic Peninsula. In order to answer outstanding questions, the workshop emphasized the need for a balanced programme of research, including investigating the oceanographic factors maintaining Antarctic sea ice, modelling ice shelves and ice streams (fast-moving glaciers), examining the geologic and recent records for analogues to test parts of the theory, and monitoring conditions in and near the ice sheet.

An important but controversial question is the possibility of a complete disappearance of the Arctic Ocean pack ice. Budyko (1969) has indicated that a 4 K temperature increase in summer might lead rather rapidly to an ice-free Arctic Ocean and there is evidence (e.g., SMIC, 1971) that this would be an irreversible process even if temperature returned to present levels. A removal of the Arctic Ocean ice could be of major climatic significance, not only over the Arctic Ocean, but also in middle and lower latitudes (Barry, 1978).

Parkinson and Kellogg (1979) have used a large-scale numerical time-dependent model of sea ice to examine the response of the Arctic Ocean ice pack to a warming of the atmosphere due to a doubling of the CO_2 concentration. The sea-ice model has been described by Parkinson and Washington (1979). The model results showed that with a 5 K surface atmospheric temperature increase the ice pack disappeared entirely in August and September but reappeared in the central Arctic Ocean in the middle of autumn. The model results were moderately dependent on assumptions regarding cloud cover. Even when atmospheric temperature increases of 6–9 K were combined with an order-of-magnitude increase in the upward heat flux from the ocean, the ice still reappeared in winter.

Manabe and Stouffer (1979, 1980) investigated how an atmospheric GCM with a coupled mixed-layer ocean with provision for a sea-ice layer responded to quadrupling the atmospheric CO_2. During the summer months the sea ice disappeared when the CO_2 was quadrupled and the change of the thermal insulation effect of the sea ice strongly influenced the seasonal variation of the warming over the Arctic region. With the present-day CO_2 concentration the model atmosphere was insulated by thicker sea ice from the influence of the underlying sea water and had a more continental climate with a larger seasonal variation of temperature than when the CO_2 was quadrupled.

Manabe and Stouffer (1979) point out that their assumption of fixed cloudiness (not computed, therefore there are no changes within a simulation or between two runs) and various simplifications contained in the sea-ice modelling mean that the quantitative results are uncertain. Similarly, Washington and Ramanathan (1980) point out that the sea-ice models have not yet been coupled with comprehensive ocean models, so that it is difficult to conclude definitively from the above studies what changes in sea-ice extent would really occur. This is an important question in view of the large-scale interactions between atmospheric circulation, sea-ice thickness and extent, and ocean circulation, which extend from the Arctic to the hemispheric or global scale.

3.6 THE EFFECT OF CO_2-INDUCED CLIMATIC CHANGES ON AGRICULTURAL PRODUCTIVITY AND TERRESTRIAL ECOSYSTEMS

The effects on agriculture of a potential CO_2-induced climatic change are a major concern. Unfortunately, these effects can only be evaluated quantitatively when reliable estimates are available for the temperature and precipitation changes in response to a CO_2 increase. It is also necessary to have models to predict the responses of individual crops to CO_2-induced climatic changes. Since all of these requirements cannot now be satisfied, it is presently possible to predict only the orders of magnitude of some of the potential effects. Kellogg and Schware (1981) have reviewed the studies that have been made of the effects on world food production; much of this section is based on the information in their review.

Kellogg and Schware first discuss the suggestion of NAS (1977b) that 'agroclimatic' zones would probably shift polewards; they point out that, although this would force farmers to adjust their agricultural practices, it may not be detrimental. If the northward penetration of the monsoon were increased, agricultural conditions in the monsoon areas might improve. A warming might also improve the agricultural productivity of subarctic regions, such as northern Russia, Canada, and Scandinavia. Kellogg and Schware quote the rule of thumb that, generally, a 1 °C increase of mean summer temperature can result in an average 10 day increase in the length of the growing season.

It is important to bear in mind that every crop responds to climatic factors differently (Kellogg and Schware, 1981; Thompson, 1975; Bach, 1978). For example, Kellogg and Schware cite these responses: a 1 °C increase in temperature leads to a 2% reduction in US corn yield (Biggs and Bartholic, 1973); under generally warmer and wetter conditions, wheat yields in Illinois and Indiana would be reduced, while wheat yields in the major producing states of Kansas, Oklahoma, North Dakota, and South Dakota would increase slightly (Ramirez *et al.*, 1975). More estimates of this type for other regions are needed. Biswas (1980) has reviewed the state of the art of crop–climate modelling.

Not only will plant physiology be affected by an increase in temperature, but pests, water availability, and soil conditions are also likely to be altered. Pimentel and Pimentel (1978) concluded that insect-pest populations will generally increase with an increase in temperature. Kellogg and Schware (1981) add that with a global warming changes would also occur in the frequency and geographical distribution of plant disease epidemics. They also point out the dangers, in view of the potential climatic changes, of the present reliance on a limited number of crop species and a narrow genetic base within each of these species.

The increase in CO_2 need not, however, have only deleterious effects. For many plant species, increased CO_2 can (AAAS, 1980):

- increase average net photosynthesis;
- change leaf area and leaf structure;
- change canopy shape;
- change the pattern of photosynthetic allocation;
- increase water use efficiency;
- increase tolerance to toxic atmospheric gases;
- change root/shoot ratios;
- change flowering dates and increase the number of flowers produced per individual;
- increase number and size of fruits and number of seeds produced per plant;
- affect germination of some species.

At present, it is not certain whether these beneficial effects on plant growth will be offset by other climate and environmental responses. The increase in average net photosynthesis due to the increased CO_2 concentration could be offset by the effects of such factors as droughts, cool temperatures, flooded soils, or lack of nutrients (NAS, 1976).

Pimentel (1980) assumes that the atmospheric CO_2 concentration would approximately double by the middle of the next century, giving rise in North America to a temperature increase of about 2 °C and an associated 10% to 30% decline in rainfall in certain regions. These climatic changes would, according to Pimentel (1980), reduce crop yields, as measured by wheat and corn, by 10% to 25%. He suggests that forest growth would also be reduced. The CO_2 fertilization effect would be more than offset by the reduced plant-growth rate.

The effects of increased CO_2 levels and of CO_2-induced climatic changes on commercial agriculture and forest and range management have been discussed in detail by AAAS (1980). For the US, there was considerable confidence in the ability of the agricultural research and development system to deliver results and products to enable farmers and foresters to adapt to changed conditions. This confidence was based on: the response to recent interannual variability; the response of the farming community and US research support to the warming trend of the 1920s to 1940s and the subsequent cooling trend; the projected capability to continue developing new varieties of the principal US crops, in view of the wide variety of conditions under which these crops are already grown.

Bach *et al.* (1981), have made a detailed analysis of food/climate interactions.

Topics covered include world food demand and supply (present and future), climatic variability and food production, vulnerability of food supplies, assessment of climate/food interactions, and policy implications of food/climate interactions. They emphasize that in the short term climatic fluctuations often provide a destabilizing influence on the year-to-year food situation that needs to be dealt with by all nations. On longer time scales, climate information can be very helpful in designing sustainable food systems within local environmental, economic, population, and other social constraints. The effects of climatic change on food production have also been considered by the National Defense University (1980). Takahashi and Yoshino (1978), and Biswas and Biswas (1979).

The result of climatic variation on crop growth has recently been investigated by Monteith (1981). After expressing his scepticism about the usefulness of multi-factorial statistical methods for describing the response of crops to weather and about computer models claiming to 'simulate' growth, he investigated the extent to which fluctuations in crop yields in Britain can be ascribed to the first-order effects of major climatic elements and attempted to predict how yields might change in response to a persistent increase in the atmospheric CO_2 concentration. He showed that light and CO_2 set an upper limit to the rate at which a crop can grow and that the duration of growth depends on temperature but may be curtailed by a lack of water. Monteith concludes that the primary effects of CO_2 would give an increase in yield of about 0.1%/year at harvest and about 7% between now and the year when the concentration reaches 400 ppm (by volume). However, he adds that the secondary effects of increases in rainfall and temperature on the scale predicted by climate models are likely to be more important for most crop yields. He suggests that a modest increase of rainfall throughout the year would benefit all crops grown in Western Europe, especially on light soils, if it were not associated with a greater risk of disease. A substantial increase of rainfall would, however, reduce yields by leaching fertilizers, reducing soil aeration, delaying the preparation of seed-beds in spring, and interfering with harvesting and hay-making. Monteith's results show that an increase of temperature would decrease cereal and potato yields in Britain and probably in much of Western Europe by about 5% per kelvin. For crops such as sugar beet and grass and for coniferous forests, the increase of photosynthesis with higher CO_2 concentrations is likely to be enhanced by higher temperatures. He concludes with the statement, 'Until the predictions from climatic models become more reliable, I see little point in developing "scenarios" for agricultural production based on numerous insecure premises. Crop models need to be improved too. . .'

Kellogg and Schware (1981) have discussed the effects of climatic change on the less managed or natural ecosystems, such as grasslands, savannahs, forests, tundra, alpine lands, and deserts. They point out that the distribution of the various biomes of the world depends primarily on climate, in particular mean temperature and precipitation, although other factors are probably involved, such as soil type and the seasonal variations of soil moisture. They conclude that, after a CO_2-induced climatic change has occurred, conditions favourable for a biome may no longer exist in its current location. The response of each ecosystem will be to seek a new equilibrium and gradually to invade neighbouring areas where the new climate is more favourable. Finally, Kellogg and Schware suggest that these shifts may be predictable for a given scenario of regional and seasonal climatic change, assuming that only natural processes are involved, but they feel that in many parts of the world human intervention will probably have more ecological impact than the climate in the next 50 to 100 years.

3.7 POSSIBLE MEASURES TO PREVENT, CONTROL OR AMELIORATE CO_2-INDUCED CLIMATIC CHANGES

A variety of measures have been proposed to prevent, control, or ameliorate CO_2-induced climatic changes. Some of them have been evaluated in some detail in the past, while others have merely been suggested. The measures include reducing the use of fossil fuels, removing CO_2 from the gases in power-plant stacks, removing CO_2 from the atmosphere or ocean, or adopting methods to cause a global cooling to counteract the potential global warming due to CO_2. At the end of this section, a few remarks on the societal aspects of control strategies are made.

(a) Reduction in the use of fossil fuels

There is considerable certainty that the observed rise in the CO_2 concentration of the atmosphere is mostly (or entirely) due to the release of CO_2 when fossil fuels are burned. Thus, if for any reason it were decided that the CO_2 concentration increase must be slowed or halted, an obvious measure would be to reduce the use of fossil fuels. Such a reduction could be brought about by reducing the energy demand and/or replacing the fossil fuels by another energy supply source (or mix).

Several studies have shown that reducing the use of fossil fuels could have a considerable effect on the atmospheric CO_2 concentration in the future. For example, Rotty (1979) showed that with a continuation of the recent 4.3% growth in the rate of use of fossil fuels until the year 2000 and a growth rate of 3% per year thereafter, the atmospheric CO_2 concentration would reach about 660 ppm (by volume) in the year 2050 and that the massive non-conventional oil production assumed did not appear significantly to enhance the global anthropogenic CO_2 production. However, with an assumed 9% per year growth in non-fossil fuel use, the use of fossil fuels would be eliminated by 2045 and the atmospheric CO_2 concentration would not increase above 500 ppm (by volume). Similarly, Figure 3.30, from Zimen (1979), shows results of calculations of the effects of different annual rates of growth of fossil-fuel use on the atmospheric CO_2 concentration. It appears that up until the year 2000 there is little difference between the CO_2 levels for the different assumed growth rates, but by the year 2040 the differences are large.

Niehaus and Williams (1979) showed the difference between two fossil-fuel strategies, in which the energy use in the year 2100 was 50 TW and 30 TW respectively. With the 50 TW fossil fuel strategy the CO_2 emissions were

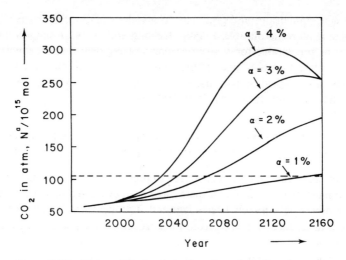

Figure 3.30 Future CO_2 levels in the atmosphere for alternative scenarios of growth in fossil fuel consumption (α is the annual growth rate). Source: Zimen (1979)

40×10^{15} g carbon per year in the year 2100 and the application of a carbon cycle model suggests that this strategy would lead to an atmospheric CO_2 concentration of 1,500 ppm (by volume) by the year 2100, five times that of the pre-industrial atmosphere. When the fossil fuel use was 30 TW in 2100, CO_2 emissions were calculated to be about 24×10^{15} g carbon per year and this strategy would lead, according to the carbon cycle model, to an atmospheric CO_2 concentration of about 1,100 ppm (by volume) in the year 2100.

The results of the carbon cycle models of all of these authors are uncertain. It certainly would not be possible to use the results as reliable predictions of the effects of the reduction of fossil-fuel use. Nevertheless, the results are probably a useful guide to the order of magnitude of these effects, particularly over the next 50 years. Thus, it can be concluded that, within this period, reducing the use of fossil fuels would certainly slow down the increase of atmospheric CO_2 concentration. Moreover, Rotty (1979b) has shown that it is not of major significance which fossil fuels are used, even though the different fossil fuels emit different amounts of CO_2 into the atmosphere per unit energy; rather, it is the overall growth rate in fossil-fuel use that plays the determining role.

Häfele (1978) has discussed reducing the use of fossil fuels by using the carbon atom more efficiently. He has pointed out that certain characteristics of fossil fuels, particularly oil, must be considered when future energy supply is discussed. Fossil fuels can be stored easily, shipped over global distances, and distributed to users in small quantities. Morever, oil plays a dominant role in supplying energy for transportation. The large-scale use of solar and nuclear energy conversion would not necessarily produce secondary energy carriers with the favourable characteristics of fossil fuels to which the present energy sector is adjusted.

Thus, Häfele (1978) suggested that solar and nuclear energy could be used to upgrade coal to produce a refined, clean, transportable fuel such as methanol. The heating value of a methanol molecule (CH_3OH) is twice as high as that of a single carbon (C) atom; the carbon atom in methanol carries with it an equal amount of energy in the form of attached hydrogen atoms. If the energy required to synthesize methanol were provided by nuclear or solar energy conversion, 1 tonne (t) of coal could roughly provide for a supply of 2 t of coal equivalent from methanol. It was also suggested that atmospheric CO_2 could be used as a source of carbon for methanol production, although this would require three times as much energy as the use of the carbon atom in coal. Häfele (1978) points out that to upgrade coal via methanol production requires investing energy and capital in substantial quantities. Capital has to be considered as the main economic substitute for presently used non-renewable resources. The capital must be invested early to build up a future long-term energy supply. For example, a supply scenario for the year 2030 presented by Häfele (1978) considers a total of 35 TW as a primary energy source, of which 15 TW are methanol. The methanol production would require the accumulation of an additional US 30×10^{12} within the next 50 years, i.e., additional to the usual capital stock formation for energy installations.

The result of using non-fossil energy sources to replace fossil fuels has also been investigated by Rotty (1979b) and Niehaus and Williams (1979). As in the cases of reducing fossil-fuel use discussed above, to replace fossil fuels by alternative sources means that the CO_2 concentration does not increase as much as with fossil-fuel strategies. A major question with regard to adopting different energy strategies is that of market penetration times, which have been discussed in detail by the Energy Systems Program (1981). In addition to the fact that market penetration is extremely regular and can be hindcast with great accuracy, another major point is the long time periods that have been required for market penetration. Häfele (1978) states that the time period for gaining or losing a 50% market share was, for the globe, 160 years for wood, 170 years for coal, and 78 years for oil, and will be 90 years for gas. For the United States, the equivalent time periods were shorter, but still in the range of 60 years. Marchetti and Nakicenovic (1979) have discussed the market penetration constraint in detail.

Questions have, however, been raised regarding the validity of the market penetration 'constraint' (e.g., Laurmann, 1980). Such questioning has implications for the risk–benefit calculations that might be made with reference to the CO_2 problem, as Laurmann (1980) has pointed out. Figure 3.31, from Laurmann (1980), shows the calculated effect of a market penetration time of 50 years on CO_2 levels. The overall energy growth is taken to be 3% per year and it is assumed that a non-fossil fuel has a 1% share of the market at $1975 + t_0$ (where t_0 can have the values −10, 0, 10, 20, or infinity). The figure indicates the near impossibility with the above assumptions, of reducing fossil-fuel use fast enough to avoid the CO_2 doubling, unless the substitute is present nuclear technology, which already has more than a 1% market share (curve $t_0 = -10$). Laurmann (1980) has concluded that a clear case for acting today to alleviate the potential CO_2 problem is difficult to make *unless* future market penetration times for new

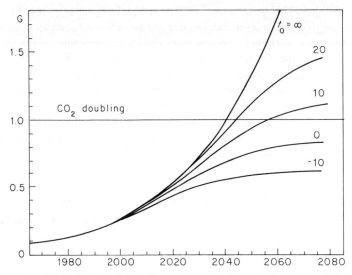

Figure 3.31   Effect of a market penetration time of 50 years on CO_2 levels, assuming a non-fossil fuel source has a 1% share of the market in the year 1975 + t_0. Source: Laurmann (1980)

energy sources can be shown to be as long as the historical times. Thus it is of importance to determine these times.

The implications of a reduction in CO_2 emissions have been discussed by Egberts and Voss (1980). They find that the reduction would mean a much more radical change in the energy supply structure than any possibly successful strategy to replace petroleum would produce. Results discussed by Egberts and Voss show that nuclear and renewable energy supplies may be able to sustain only about 50% of the growth of primary energy demand in the countries belonging to the International Energy Agency. They conclude that reductions of CO_2 emissions can be achieved only by huge efforts in energy saving, connected with a full utilization of nuclear and renewable resources.

(b) The collection, disposal, and storage of CO_2

Several studies have examined whether it would be possible to collect the CO_2 released from fossil fuels, convert it to an appropriate form for transport and disposal and then store it. Baes et al. (1980) have discussed the technical feasibility, monetary cost, and energy cost of the options that appear the most promising. They point out that Marchetti (1977) first gave serious consideration to methods of collection and disposal, suggesting that liquid CO_2 could be piped from power-plant scrubbers to injection points in the ocean where the water flows down to intermediate or deep layers, in the Straits of Gibraltar, for example. Baes et al. (1980) have considered primarily collection from concentrated sources. They suggest that, although power plants account for only one-third of the fossil-fuel CO_2 released globally and only one-sixth of that released in the United States, such sources are likely to become the major contributors in the future.

Baes *et al.* found that scrubbing CO_2 from the flue gas of a coal-fired power station by the most efficient method presently available would require about 43% of the combustion energy of the coal. They concluded that simultaneous CO_2–SO_2 scrubbing does not appear practical because of the great difference in the concentrations of the two gases. They quote also the suggestion made by Marchetti (1978) that the coal should be burned in pure oxygen to avoid the scrubbing step. The oxygen supply would require at least 30% of the combustion energy.

The best option for disposal of CO_2 is, according to Baes *et al.*, disposal in the deep ocean where the low rate of circulation should give a retention time of hundreds of years. They suggested that the CO_2 could be injected as a concentrated sea water solution, or in the form of liquid CO_2, blocks of solid CO_2 hydrate, or dry ice. They pointed out, however, that sites for disposal in the ocean that are near the shore and energy markets and where the water is sufficiently deep are not abundant. They concluded that the most efficient option for the future might be floating, coal-fired power plants that use cold deep sea water for both condenser cooling and CO_2 disposal. As a near-term option, they suggested that depleted oil and gas fields could be used for disposal of CO_2. Storage near the South Pole or as buried biomass were considered to be the least attractive disposal alternatives.

Steinberg and Albanese (1980) have also looked at the technological aspects of CO_2 removal and disposal. They found that removing CO_2 from the atmosphere or the surface waters of the oceans is not feasible if a fossil energy source is used to drive the system, since more CO_2 would be generated than recovered. They also concluded that, for 60% to 90% absorption of the CO_2 contained in the flue gas of a coal-fired power station using sea water and deep ocean disposal, more CO_2 would be generated in supplying energy to the control system than would be recovered from the flue gas. Removal and recovery of CO_2 from the stacks of fossil-fuel plants by commercially available processes was considered to be feasible. However, the authors emphasized that these processes require large amounts of energy and thus reduce the efficiency of power generation. For example, for 90% CO_2 removal the efficiency decreases from 34% to between 15% and 6%. For the disposal of the CO_2, Steinberg and Albanese find that deep ocean disposal is a possibility, but point out that the costs are substantial as the effects of injected CO_2 on the ocean ecosystem are unknown. The costs of gaseous and liquid pipeline disposal were found to be much greater than for solid disposal. Steinberg and Albanese conclude that for effective control of atmospheric CO_2 by removal from flue gases, it would be advantageous and perhaps necessary to centralize the use of fossil fuels.

Mustacchi *et al.* (1978) have discussed disposing of CO_2 from power plants and suggest three modes of disposing of CO_2 into the ocean: by bubbling and dissolution of untreated flue gases; by dissolution of separated CO_2 in the gaseous or liquid phase; and by dissolving the CO_2 in sea water in a pressurized absorber and pumping the solution back into the ocean. In the case of CO_2 separation and disposal in the gaseous phase, the plant would consist of a cooling tower to cool the flue gases down to 40 °C and an absorber–stripper system. Solid strippers were evaluated by Mustacchi et al. (1978) and found unsuitable; of the numerous liquid strippers, amines were found to be the best. Using this process, it was suggested that the additional cost per kWh of electricity would not be more than 20%.

A further option for CO_2 control is to recover the CO_2 removed from the stack gases (or atmosphere or ocean) and reuse it or recycle it. Recovery would require treating the solids or liquids used for CO_2 removal, e.g., desorption by flashing or distillation, stripping by gases, or decomposition (Steinberg et al., 1978). In some cases, however, it may be possible to proceed directly from CO_2 removal to CO_2 reuse. Steinberg et al. (1978) point out that, although a number of CO_2-containing commercial products could involve the reuse of recovered CO_2, e.g., dry ice, soda ash, inorganic carbonates, or urea, the atmospheric CO_2 content would remain the same if fossil fuels were used to produce them; in any case the markets for the products are limited.

The most discussed approach for using recovered carbon is to convert it into liquid or gaseous fuels using a non-fossil energy source.

For example, as mentioned in Section 3.7(a), Häfele (1978) has suggested using CO_2 recovered from the atmosphere with hydrogen produced by nuclear or solar energy to produce methanol. Steinberg et al. (1978) likewise suggest that the recovered CO_2 could be combined with electrolytically produced hydrogen to form synthetic carbonaceous fuels. They show, however, that there is more than enough CO_2 emitted from power plants using fossil fuels to supply the needs of the automotive industry in the US, but that about a thousand 1,000 MW(e) nuclear power plants would be required to make the total conversion. Further possibilities for reusing CO_2 include the tertiary treatment of oil wells, which may be of limited value as a long-term disposal method (Steinberg et al., 1978), and using CO_2 as a diluent for oxygen in a fossil-fuel power plant so that the only effluent is CO_2.

Other suggestions have been made for the removal and storage of CO_2. For example, Marchetti (1979) discusses a scheme proposed by Dyson (1977), which involves planting trees, which would remove CO_2 from the atmosphere and storage of it in wood. The proposal would involve planting 10^{12} sycamores, but Marchetti (1979) points out some drawbacks. Firstly, the scale of the operation is 'almost unthinkable' in terms of present technology and social organization. Secondly, the forests have a low albedo and thus absorb more solar radiation than most other natural or artificial surfaces. Marchetti (1979) calculates that, with a change of albedo of 20% caused by trees, the forest would have to accumulate 20 kg of CO_2 (equivalent) per square metre, so that the increased warming due to the absorption by the forest and the decrease in the warming due to the removal of CO_2 from the atmosphere are balanced. Thus it is suggested that all of the trees would have to be planted now in order that they would be working in a cooling mode by the end of the century.

(c) Altering climatic boundary conditions to counteract a CO_2-induced climatic change

Some suggestions have been made along the lines of counteracting CO_2-induced climatic changes by other kinds of interference. For example, Munn and Machta (1979) refer to suggestions for adding dust or aerosols over the ocean, or reducing the atmospheric concentration of another greenhouse gas with a much smaller content that might be easier to remove.

Another alternative that has been proposed, but that is not justified physically, is that the net radiative balance with increased CO_2 might be returned to pre-industrial values by increasing the ground albedo. One suggestion involved distributing small pieces of reflective material over the ocean surface or in the stratosphere, in particular, small pieces of latex sheet (0.01 mm thick), of which 10 t/km^2 would be required. However, it has been pointed out (Bach, 1979) that the extremely high costs and the risk of further environmental damage, even if only 1% of the earth's surface were treated in this way, are disadvantageous.

Another suggestion considers the large-scale deployment of solar-energy conversion systems and the possibility that the albedo change from them can be used to compensate for the CO_2-induced warming (Bach, 1979). Preliminary calculations made by Bach and Schwanhäußer (1978) suggested that it would be possible to deploy solar energy conversion systems such that the albedo change compensates for the CO_2-induced warming. However, more detailed studies, such as that of Bhumralkar et al. (1979) disagree with this conclusion by showing that, for example, although they make a large reflective surface, the mirrors collect solar energy by focusing it and the overall effect is that of adding heat to the climate system. Likewise, photovoltaic cell arrays would probably cause negligible albedo changes if placed in naturally dark surface (e.g., forested) areas and would cause a decrease of albedo (warming) if placed in desert areas. Large energy plantations might also be expected to cause a decrease in albedo, particularly since forests have a lower albedo than most other natural surfaces.

Most of the suggestions for manipulating climatic boundary conditions to counteract a CO_2-induced warming are speculative and their engineering feasibility has not been thought out (Munn and Machta, 1979). Moreover, as discussed thoroughly by Kellogg and Schneider (1974), there are many problems associated with the concept of climate modification. The first reason for this is that we are unable to predict at the present time what the effects of any large-scale modification scheme would be. As pointed out by Kellogg and Schneider (1974), it would be extremely irresponsible to tamper with a system that is a major determinant in the livelihood and lifestyles of the world's population; on the other hand, those who live in affected regions, where a prediction of a local disaster from a climate change could be a forecast, might not agree that an offsetting change would be irresponsible.

Global climatic changes are to be expected in the form of large regional anomalies. Although increased atmospheric CO_2 might cause a warming in global terms, different regions will experience warming or cooling of varying magnitudes along with regional rainfall anomalies. In view of the non-linearity, it is difficult to imagine that a simple countervailing measure, such as increasing the global albedo could counteract specific regional anomalies. Moreover, since many of the feedbacks within the climate system are not completely understood, the impact of countervailing measures can also not be understood or predicted. Therefore, any such measures could only be evaluated realistically after a more complete knowledge of the climate system is available.

(d) Further general remarks on control strategies

Even in the absence of an ability to predict the specific climatic changes that will occur as a result of energy systems, assessments are necessary if the world is to prepare itself to devise and implement effective solutions to the problems that may arise. It has been suggested (AIHS, 1978) that a critical variable will be society's choice of whether to adapt to a climatic change or whether to forestall or limit it.

The Aspen Workshop (AIHS, 1978) has pointed out that certain general issues will arise regardless of whether society chooses the control or adaptation strategy. These include the following:

1. *Resource-mobilization issues.* Whatever strategy is selected, mobilizing large economic and human resources would be required. For example, adaptation would require significant expenditures for water-resources development, agricultural research, etc. Some of the costs of control have been presented earlier in this section.
2. *Distributing losses and gains.* Society may have to distribute losses and gains as a consequence of climatic change and there will be considerable debate as to what extent losses of specific groups of people should be borne by all of mankind.
3. *Altered structure of society.* Climatic change could also bring about altered distributions of influence and power. Changes in energy supply systems could also have an impact on political and economic systems, as could changes in food and water resources.

These issues involve equity and the ability to influence decisions by different groups, who may view costs and benefits differently (Baes *et al.*, 1976). This is especially so since costs and benefits will not be distributed evenly over different regions. Wiser (1978) has also highlighted some of these issues and asks, for example, these questions: If we were able to establish the amount of global reduction of atmospheric CO_2 required, how is the reduced CO_2 emission to be divided among nations? Can, or should, a nation that stands to gain by a CO_2-induced climatic change be forced to forego these gains in order to 'protect' the rest of the world? These points might also apply to other climatic changes or other energy systems.

Meyer-Abich (1980) suggests that basically three political responses are possible to a CO_2-induced climatic change:

- prevention of additional CO_2 generation;
- compensation by suspending undesirable effects by global efforts (international activities and budgets);
- adaptation by responding to undesirable effects of climatic change on the national or individual level.

Examining each of these responses in turn, Meyer-Abich finds that compensation and prevention are much less practical from a political point of view. Adaptation, on the other hand, means that expenses are deferred most distantly into the future and has the advantage that long-term international cooperation or agreement on long-term goals is not required. Thus, Meyer-Abich concludes that adaptation is presently the most rational political action. However, he points out that the conflict potential between North and South could be considerably enhanced by climate changes, because the already existing inequalities in distribution of wealth may tend to be increased. If it is mainly the developing countries that are affected, then, Meyer-Abich concludes, the climate-oriented policies become part of development policies in general. Compared with the already existing problems in development policy, the possibility of climatic changes, therefore, is a 'marginal' problem, since it is not qualitatively different and does not significantly increase the tensions existing already. Thus, Meyer-Abich finds no problems that can be uniquely attributed to CO_2-induced climatic change and suggests that all of the problems associated with climate change should be addressed at least as much for other reasons.

Social and institutional responses to the CO_2 issue were discussed in detail at a recent workshop (AAAS, 1980; pp. 79–103). It was pointed out that the CO_2 issue is a gradually developing problem that is so far proceeding too slowly to attract significant public notice. However, there are aspects linking it to other high priority social problems, including the development of alternative energy systems and certain environmental threats. The workshop emphasizes that at present uncertainties inhibit precise definition of the social costs and benefits of CO_2-induced climatic change.

In an analysis of the economic and geopolitical consequences of a CO_2-induced climatic change at the same workshop (AAAS, 1980, pp. 104–120), it was emphasized that one must keep the potential effects in perspective with the huge social, political, and physical or natural events that are likely to occur over the next 50 to 100 years. That is, other events are likely (but not certain) to have effects of comparable or greater magnitude than (and will interact with) the CO_2 effects. It was concluded that evaluation of the effects of possible CO_2-induced climatic changes should begin now, even though the geophysical models are unable to specify details of future climate.

3.8   ASSESSMENT OF RISKS ASSOCIATED WITH CO_2-INDUCED CLIMATIC CHANGES

Each of the energy-supply alternatives, and therefore any mix of them, has particular risks for the environment, human health, and safety. Therefore, any decision on energy supply alternatives involves making trade-offs between different types of risks and hazards. Thus, one must define what is 'better' or 'worse' in an environmental sense and measure it against other costs and benefits of energy systems, including economic growth, availability and reliability of supply, price of production, and use of energy. Again the problem is to determine a socially acceptable trade-off among the environmental and non-environmental objectives (Energy Systems Program, 1981). As will be shown in this section, this task has barely begun.

Available studies of the risks associated with energy systems have concentrated on the risks of climatic change due to CO_2 released from fossil-fuel combustion. Laurmann (1978) examined the uncertainties in the risk evaluation, for example, the uncertainties in the predicted rate of increase of CO_2 and in the associated climatic change. Figure 3.32 from Laurmann (1978) shows the effect on global temperature increase of changing the atmospheric CO_2 growth rates. The temperature increase was calculated using a formula based on the results of climate models, with the assumption that this relation could be in error by a factor of two. The figure shows that a 1% change from a 3% per year CO_2 growth rate shifts the time to elevate temperatures by 2.5 °C by about a decade. Laurmann (1978) discusses the uncertainties due to inadequacies in climate modelling by showing the cumulative probability of reaching at least a 2.5 °C global temperature increase by a given date, assuming that the curves shown in Figure 3.32 have a log normal error of 2. Laurmann (1978) concludes that the most likely date for reaching a 2.5 °C global temperature increase is the year 2020 but that this date may be either considerably advanced or may never occur. Thus, he suggests using the probabilistic techniques of decision analysis, which would have

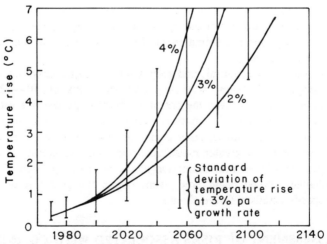

Figure 3.32   Global temperature increase computed for various atmospheric CO_2 concentration growth rates. Source: Laurmann (1978)

to include discounting future risks and benefits. An example is given of how discounting reduces the time taken to encounter maximum costs.

A study of the effect of a CO_2-induced warming has been made by Schneider and Chen (1979), who have investigated the implications of a 5–8 m rise in ocean level for US coastal states. Such a rise could come about if the West Antarctic ice sheet were to surge. Schneider and Chen calculated that over 11 million people (5.7% of the continental US population) plus some US $110,000 million (1971 value) in non-removable property value would be affected by a 5 m rise and about 16 million people and US $150,000 million (1971 value) for an 8 m rise. On a regional scale the effects were found to be more severe: some 40% of Florida's population and 50% of its income and immobile wealth would be affected by a 5 m rise.

Schneider and Chen point out that the time frame for such a sea-level change would be crucial. For instance, if the rise were to occur rapidly over a few years' time with little or no prior warning, an event that is extremely unlikely, then the logistic problems of housing and feeding people and of salvaging personal property, equipment, and inventories would be enormous. A slower rise could be more easily accommodated.

Schneider (1979) has extended this study to illustrate the effect of discounting. Firstly, the estimate of the real property costs to the United States for a 8 m sea-level rise were generalized to the rest of the world by assuming that the total world value (in 1972 US dollars) of inundated areas is 10 times greater than the US losses (i.e., total world value is 2.5×10^{12}). It was also assumed that the losses from such flooding occur in 150 years. One can then ask: What is it worth to us today to invest to prevent such a loss in 150 years? To answer this, Schneider assumes that the future can be discounted at 7% per year, which means that a dollar in 150 years time is worth only about 1/30,000 of a dollar today. Thus a 2.5×10^{12} inundation cost in 150 years is worth only about $75 million, which is considerably less than the economic value of fossil-fuel-related industries or the potential agricultural income from deforested lands. Thus, Schneider (1979) concludes that many people today would not consider it economically rational to spend $75 million now even to prevent a 2.5×10^{12} dollar catastrophe 150 years in the future. This assumes, as Schneider points out, that the 7% discount rate is economically rational. If the inundation were to occur in only 20 years, it would cost 6×10^{11}, a loss sufficiently serious to warrant immediate policy actions. Uncertainties in the timing of potential sea-level rises combine with those of estimating the discount rate, to make assessment of such an effect difficult. However, Schneider concludes that discounting the future at high rates further diminishes the likelihood that this generation will invest heavily to hedge against potential CO_2-induced losses in the future.

Cost–benefit assessments related to the potential climatic changes due to energy systems are presently of a very preliminary nature. Emphasis has been placed so far on studies related to the potential increase of atmospheric CO_2 concentration, since this has been generally considered to represent the most immediate 'threat'. In the terms of risk assessment, the first step is to define the probability quantitatively that a certain climate change will occur due to the CO_2 increase and the consequence of this climatic change. The uncertainties that must be attached to these estimates must also be evaluated. In terms of any policy action that might ultimately be based on risk assessment, a major point is that the CO_2 problem is a global problem, which will manifest itself differently in different regions, such that many of the issues that Schneider (1980) has discussed will play a major role, e.g., risk heterogeneity, equity, and risk intercomparison.

Very many questions must be answered quantitatively before analyses can be made of the costs of particular energy strategies and of any 'remedial' actions defined. In particular, the climatic effect of energy systems must be defined quantitatively on a regional basis and the costs of such effects must be evaluated.

REFERENCES

AAAS (1980). Workshop on environmental and societal consequences of a possible CO_2-induced climate change. CONF-7904143, United States Department of Energy, Washington, DC.

AIHS (1978). The consequences of a hypothetical world climate scenario based on an assumed global warming due to increased carbon dioxide. *Proc. of Symposium Aspen Institute for Humanistic Studies*. Aspen, Colorado.

Armentano, T. V. (ed.) (1980). The role of organic soils in the world carbon cycle—Problem analysis and research needs. CONF-7905135, US Department of Energy, Washington, DC.

Armentano, T. V., and J. Hett (eds.) (1980). The role of temperate zone forests in the world carbon cycle—Problem definition and research needs. CONF-7903105 UC-11, US Department of Energy, Washington, DC.

Augustsson, T., and V. Ramanathan (1977). A radiative–convective model study of the CO_2 climate problem. *J. Atmos. Sci.*, 448–451.

Ausubel, J. H. (1980). Climatic change and the carbon wealth of nations. WP-80-75, International Institute for Applied Systems Analysis, Laxenburg, Austria.

Bacastow, R. B. (1976). Modulation of atmospheric carbon dioxide by the southern oscillation. *Nature*, **261**, 116.

Bacastow, R. B., and C. D. Keeling (1973). Atmospheric carbon dioxide and radiocarbon in the natural carbon cycle: II. Changes from A.D. 1700 to 2070 as deduced from a geochemical model. In, G. M. Woodwell and E. V. Pecan (eds.), *Carbon and the Biosphere*. USAEC, Springfield, Virginia.

Bacastow, R. B., and C. D. Keeling (1979). Models to predict future atmospheric CO_2 concentrations. In, W. P. Elliott and L. Machta (eds.), Workshop on the global effects of carbon dioxide from fossil fuels. CONF-770385, UC-11, US Department of Energy, Washington, DC.

Bach, W. (1978). The potential consequences of increasing CO_2 levels in the atmosphere. In, J. Williams (ed.), *Carbon Dioxide, Climate and Society*. Pergamon Press, Oxford, England.

Bach, W. (1979). Untersuchung der Beeinflussung des Klimas durch anthropogene Faktoren. Report to the Umweltbundesamt. University of Münster, FR Germany.

Bach, W., J. Pankrath, and W. W. Kellogg (eds.) (1979). *Man's Impact on Climate*, Elsevier, Amsterdam.

Bach, W., J. Pankrath, and S. H. Schneider (eds.) (1981). *Food–Climate Interactions*. Reidel, Dordrecht, Holland.

Bach, W., J. Pankrath, and J. Williams (eds.) (1980). *Interactions of Energy and Climate*. Reidel, Dordrecht, Holland.

Bach, W., and G. Schwanhäußer (1978). Can the right energy mix prevent climatic change? *Proc. of Conference on Climate and Energy: Climatological Aspects and Industrial Operations.* American Meteorological Society. Boston, Massachusetts.

Baes, C. F., Jr., H. E. Goeller, J. S. Olson, and R. M. Rotty (1976). The global carbon dioxide problem. ORNL-5194, Oak Ridge National Laboratory, 78 pp.

Baes, C. F., Jr., S. E. Beall, D. W. Lee, and G. Marland (1980). The collection, disposal, and storage of carbon dioxide. In, W. Bach, J. Pankrath, and J. Williams (eds.), *Interactions of Energy and Climate.* Reidel, Dordrecht, Holland.

Barry, R. G. (1978). Cryospheric responses to a global temperature increase. In, J. Williams (ed.), *Carbon Dioxide, Climate and Society*, Pergamon Press, Oxford, England.

Baumgartner, A. (1979). Climate variability and forestry. *Proceedings of the World Climate Conference.* WMO Publication No. 537, World Meteorological Organization, Geneva.

Bernard, H. W., Jr. (1980). *The Greenhouse Effect.* Ballinger, Cambridge, Massachusetts.

Bhumralkar, C. M., J. Williams, and A. Slemmons (1979). The impact of a conceptual solar thermal electric conversion plant on regional meteorological conditions: A numerical study. *Solar Energy*, 23, 393–403.

Biggs, R. H., and J. F. Bartholic (1973). Agronomic effects of climate change. In, A. J. Broderick (ed.), *Proceedings of the Second Conference on CIAP.* US Department of Transportation, Washington, DC.

Biswas, A. K. (1980). Crop–climate models: A review of the state of the art. In, J. Ausubel and A. K. Biswas (eds.), *Climatic Constraints and Human Activities.* Pergamon Press, Oxford, England.

Biswas, M. R., and A. K. Biswas (eds.) (1979). *Food, Climate and Man.* John Wiley and Sons, New York.

Björkström, A. (1979). A model of the CO_2 interaction between atmosphere, oceans, and land biota. In, B. Bolin *et al.* (eds.), *The Global Carbon Cycle.* Wiley, New York.

Bolin, B. (1979). Global ecology and man. *Proceedings of the World Climate Conference.* WMO Publication No. 537. World Meteorological Organization, Geneva.

Bolin, B., E. T. Degens, P. Duvigneaud, and S. Kempe (1979). The global biogeochemical carbon cycle. In, B. Bolin *et al.* (eds.), *The Global Carbon Cycle.* Wiley, New York.

Broecker, W. S. (1977). The fate of fossil fuel CO_2: A research strategy. In, Report of the scientific workshop on atmospheric carbon dioxide, WMO-474, World Meteorological Organization, Geneva.

Broecker, W. S., T. Takahashi, H. J. Simpson, and T.-H. Peng (1979). Fate of fossil fuel carbon dioxide and the global carbon budget. *Science*, 206, 409–418.

Brown, S., and A. E. Lugo (1980). Preliminary estimate of the storage of organic carbon in tropical forest ecosystems. In, S. Brown *et al.* (eds.). The role of tropical forests on the world carbon cycle. CONF-800350, US Department of Energy, Washington, DC.

Brown, S., A. E. Lugo, and B. Liegel (1980). The role of tropical forests on the world carbon cycle. CONF-800350, US Department of Energy, Washington, DC.

Budyko, M. I. (1969). The effect of solar radiation variations on the climate of the earth. *Tellus*, 21, 611–619.

Callendar, G. S. (1958). On the amount of carbon dioxide in the atmosphere. *Tellus*, 10, 243–248.

Chan, Y.-H., J. S. Olson, and W. R. Emanuel (1980). Land use and energy scenarios affecting the global carbon cycle. *Environmental International*, 4, 189–206.

CIAP (1975). Impacts of climatic change on the biosphere. CIAP Monograph 5, Part 2—*Climatic Effects.* Climatic Impacts Assessment Program. Department of Transportation, Washington DC.

Clark, W. C. (1982). *Carbon Dioxide Review 1982.* Oxford University Press, Oxford, England.

Dyson, F. J. (1977). Can we control the carbon dioxide in the atmosphere? *Energy*, **2**, 287–291.

Eckholm, E. (1979). Planting for the future: Forestry for human needs. Worldwatch Institute, Washington, DC.

Egberts, G., and A. Voss (1980). Reduction of fossil fuel use and adoption of alternative energy sources. In, W. Bach, J. Pankrath, and J. Williams (eds.), *Interactions of Energy and Climate*. Reidel, Dordrecht, Holland.

Elliott, W. P., and L. Machta (eds.) (1979). Workshop on the global effects of carbon dioxide from fossil fuels. CONF-770385, US Department of Energy, Washington, DC.

Energy Systems Program (1981). *Energy in a Finite World: Volume 1. Paths to a Sustainable Future; Volume 2. A Global Systems Analysis*. Ballinger Publishing Company, Cambridge, Massachusetts.

Flohn, H. (1980). Possible climatic consequences of a man-made global warming. International Institute for Applied Systems Analysis, Laxenburg, Austria.

Gates, W. L., and K. H. Cook (1980). Preliminary analysis of experiments on the climatic effects of increased CO_2 with the OSU atmospheric general circulation model. Report No. 14, Climatic Research Institute, Oregon State University, Corvallis, Oregon.

GISS (1978). Proposal for research in global carbon dioxide source/sink budget and climate effects. Goddard Institute for Space Studies, New York (reported in NAS, 1979).

Häfele, W. (1978). A perspective on energy systems and carbon dioxide. In, J. Williams (ed.), *Carbon Dioxide, Climate and Society*. Pergamon Press, Oxford, England.

Hampicke, U. (1979). Sources and sinks of carbon dioxide in terrestrial ecosystems. (Submitted to *Environmental International*.)

Hampicke, U. (1980). The role of the biosphere. In, W. Bach, J. Pankrath, and J. Williams (eds.), *Interactions of Energy and Climate*. Reidel, Dordrecht, Holland.

Hansen, J., D. Johnson, A. Lacis, S. Lebedeff, P. Lee, D. Rind, and G. Russell (1981). Climate impact of increasing atmospheric carbon dioxide. *Science*, **213**, 957–966.

Hansen, J. E. (1980). Private communication (reported in NAS, 1979; in preparation for *J. Atmos. Sci.*).

Hasselmann, K. (1979). On the problem of multiple time scales in climate modeling. In, W. Bach *et al.* (eds.), *Man's Impact on Climate*. Elsevier, Amsterdam.

Hoffert, M. I., A. J. Callegari, and C.-T. Hsieh (1980). The role of deep sea heat storage in the secular response to climatic forcing. *J. Geophys. Res.*, **85**, 6667–6679.

Hollin, J. T. (1970). Interglacial climates and Antarctic ice surges. *Quat. Res.* **2**, 401–408.

Hughes, T. (1975). The West Antarctic ice sheet: Instability, disintegration and initiation of ice ages. *Rev. Geophys. Space Phys.*, **13**, 502–526.

Hunt, B. G., and N. C. Wells (1979). An assessment of the possible future climatic impact of carbon dioxide increases based on a coupled one-dimensional atmospheric–ocean model. *J. Geophys. Res.*, **84**, 787–791.

Idso, S. B. (1980). The climatological significance of a doubling of earth's atmospheric carbon dioxide concentration. *Science*, **207**, 1462–1463.

Keeling, C. D. (1973). The carbon dioxide cycle: Reservoir models to depict the exchange of atmospheric carbon dioxide with the oceans and land plants. In, S. I. Rasool (ed.), *Chemistry of the Lower Atmosphere*. Plenum Press, New York, pp. 251–329.

Keeling, C. D. (1980). The oceans and biosphere as future sinks for fossil fuel carbon dioxide. In, W. Bach, J. Pankrath, and J. Williams (eds.), *Interactions of Energy and Climate*. Reidel, Dordrecht, Holland.

Keeling, C. D., and R. Bacastow (1977). Impact of industrial gases on climate. In, *Energy and Climate*. National Academy of Sciences, Washington, DC.

Keeling, C. D., R. B. Bacastow, A. E. Bainbridge, C. A. Ekdahl, P. R. Guenther, and L. S. Waterman (1976). Atmospheric carbon dioxide variations at Mauna Loa Observatory, Hawaii. *Tellus*, **28**, 538–551.

Kellogg, W. W. (1977). Effects of human activities on global climate. Technical Note No. 156, WMO—No. 486, World Meteorological Organization, Geneva, Switzerland.

Kellogg, W. W., and S. H. Schneider (1974). Climate stabilization: For better or for worse. *Science*, **186**, 1163–1172.

Kellogg, W. W., and R. Schware (1981). *Climate Change and Society: Consequences of Increasing Carbon Dioxide*. Westview Press, Boulder, Colorado.

Killough, G. G., and W. R. Emanuel (1981). A comparison of several models of carbon turnover in the ocean with respect to their distributions of transit time and age, and responses to atmospheric CO_2 and ^{14}C. *Tellus*, **33**, 274–290.

Kukla, G., and J. Gavin (1981). Summer ice and carbon dioxide. *Science*, **214**, 497–503.

Lamb, H. H. (1972). Climate: *Present, Past and Future. Vol. 1, Fundamentals and Climate Now*. Methuen, London.

Lamb, H. H. (1977). *Climate, Present, Past and Future. Vol. 2, Climatic History and the Future*. Methuen, London.

Laurmann, J. A. (1978). Fossil fuel utilization policy assessment and CO_2 induced climatic change. In, J. Williams (ed.), *Carbon Dioxide, Climate and Society*. Pergamon Press, Oxford, England.

Laurmann, J. A. (1980). Assessing the importance of CO_2-induced climatic changes using risk benefit analysis. In, W. Bach, J. Pankrath, and J. Williams (eds.), *Interactions of Energy and Climate*. Reidel, Dordrecht, Holland.

Lian, M. S., and R. D. Cess, (1977). Energy balance climate models: A reappraisal of ice-albedo feedback. *J. Atmos. Sci.*, **34**, 1058–1062.

Lugo, A. E. (1980). Are tropical forest ecosystems sources or sinks of carbon? In, S. Brown *et al.* (eds.), The role of tropical forests on the world carbon cycle. CONF-800350, US Department of Energy, Washington, DC.

MacCracken, M. C. (1980). Climate research. Carbon Dioxide Research Progress Report, DOE-EV-0071, US Department of Energy, Washington, DC.

Machta, L. (1973). Prediction of CO_2 in the atmosphere. In G. M. Woodwell and C. V. Pecan (eds.), *Carbon and the Biosphere*. USAEC, Springfield, Virginia.

Machta, L. (1979). Atmospheric measurements of carbon dioxide. In, W. P. Elliott and L. Machta (eds.), Workshop on the global effects of carbon dioxide from fossil fuels. CONF-770385 UC-11, US Department of Energy, Washington, DC.

Machta, L., and W. P. Elliott (1980). Carbon cycle research. Carbon Dioxide Research Progress Report, DOE-EV-0071, US Department of Energy, Washington, DC.

Madden, R. A., and V. Ramanathan (1980). Detecting climate change due to increasing carbon dioxide. *Science*, **209**, 763–768.

Manabe, S., K. Bryan, and M. J. Spelman (1979). A global ocean–atmosphere climate model with season variation for future studies of climate sensitivity. *Dyn. Atmos. Oceans*, **3**, 393–426.

Manabe, S., and R. J. Stouffer (1979). A CO_2 climate sensitivity study with a mathematical model of the global climate. *Nature*, **282**, 491–493.

Manabe, S., and R. J. Stouffer (1980). Sensitivity of a global climate model to an increase of CO_2 concentration in the atmosphere. *J. Geophys. Res.*, **85**, 5529–5554.

Manabe, S., and R. T. Wetherald (1967). Thermal equilibrium of the atmosphere with a given distribution of relative humidity. *J. Atmos. Sci.*, **24**, 241–259.

Manabe, S., and R. T. Wetherald (1975). The effects of doubling the CO_2 concentration on the climate of a general circulation model. *J. Atmos. Sci.*, **32**, 3–15.

Manabe, S., and R. T. Wetherald (1980). On the distribution of climate change resulting from an increase in CO_2 content of the atmosphere. *J. Atmos. Sci.*, **37**, 99–118.

Marchetti, C. (1977). On geoengineering and the CO_2 problem. *Climatic Change*. **1**. 59–68.

Marchetti, C. (1978). Constructive solutions to the CO_2 problem. International Institute for Applied Systems Analysis, Laxenburg, Austria. Unpublished manuscript.

Marchetti, C. (1979). Constructive solutions to the CO_2 problem. in, W. Bach et al., *Man's Impact on Climate*. Elsevier, Amsterdam.

Marchetti, C., and N. Nakicenovic (1979). The dynamics of energy systems and the logistic substitution model. RR-79-13, International Institute for Applied Systems Analysis, Laxenburg, Austria.

Mercer, J. H. (1978). West Antarctic ice sheet and CO_2 greenhouse effect: A threat of disaster. *Nature*, **271**, 321–325.

Meyer-Abich, K. M. (1980). Chalk on the white wall? On the transformation of climatological facts into political facts. In, J. Ausubel and A. K. Biswas (eds.), *Climatic Constraints and Human Activities*. Pergamon Press, Oxford, England.

Miller, P. C. (ed.) (1980). Carbon balance in northern ecosystems and the potential effect of carbon-dioxide-induced climatic change. CONF-8003118, US Department of Energy, Washington, DC.

Mitchell, J. M. (1979). Some considerations of climatic variability in the context of future CO_2 effects on global-scale climate. In, W. P. Elliott and L. Machta (eds.), Workshop on the global effects of carbon dioxide from fossil fuels. CONF-770385 US-11, US Department of Energy, Washington, DC.

Monteith, J. L. (1981). Climatic variation and the growth of crops. *Quart. J. Roy. Meteor. Soc.*, **107**, 749–774.

Munn, R. E., and L. Machta (1979). Human activities that affect climate. *Proc. of the World Climate Conference*. WMO, Publication No. 537, World Meteorological Organization, Geneva.

Mustacchi, C., P. Armenante, and V. Cena (1978) Carbon dioxide disposal in the ocean. In, J. Williams (ed.), *Carbon Dioxide, Climate and Society*. Pergamon Press, Oxford, England.

NAS (1976). Climate and food: Climatic fluctuation and US agricultural production. Committee on Climate and Weather Fluctuations and Agricultural Production, National Academy of Sciences, Washington, DC.

NAS (1977a). *Energy and Climate*. National Academy of Sciences, Washington, DC.

NAS (1977b). *Food and Nutrition Study*. National Academy of Sciences, Washington, DC.

NAS (1979). *Carbon Dioxide and Climate: A Scientific Assessment*. National Academy of Sciences, Washington, DC.

National Defense University (1980). *Crop Yields and Climatic Change to the Year 2000*, Vol. 1. Ft. Leslie McNair, Washington, DC.

Newell, R. E., and T. G. Dopplick (1979). Questions concerning the possible influence of anthropogenic CO_2 on atmospheric temperature. *J. Appl. Meteor.*, **18**, 822–825.

Niedercorn, J. H. (1976). The capital costs of climatically induced shifts: the example of the American corn belt. In, T. A. Farrar (ed.), *The Urban Costs of Climate Modification*. Wiley, New York.

Niehaus, F. (1976). A non-linear tandem model to calculate the future CO_2 and ^{14}C burden to the atmosphere. RM-76-35, International Institute for Applied Systems Analysis, Laxenburg, Austria.

Niehaus, F., and J. Williams (1979) Studies of different energy strategies in terms of their effects on the atmospheric CO_2 concentration. *J. Geophys. Res.*, **84**, 3123–3129.

Nordhaus, W. D. (1976). Economic growth and climate; the carbon dioxide problem. Annual Meeting Amer. Econ. Assoc.

Nordhaus, W. D. (1977). Strategies for the control of carbon dioxide. Cowles Foundation Discussion Paper No. 443.

Oeschger, U., U. Siegenthaler, U. Schotterer, and A. Gugelmann (1975). A box diffusion model to study the carbon dioxide exchange in nature. *Tellus*, **27**, 168–192.

Ohring, G., and P. Clapp (1980). The effect of changes in cloud amount on the net radiation at the top of the atmosphere. *J. Atmos. Sci.*, **37**, 447–454.

Olson, J. S., H. A. Pfuderer, and Y.-H. Chan (1978). Changes in the global carbon cycle and the biosphere. ORNL/EIS-109, Oak Ridge National Laboratory, Oak Ridge, Tennessee.

Oort, A. H., and T. H. Vonder Haar (1976). On the observed annual cycle in the ocean–atmosphere heat balance over the northern hemisphere. *J. Phys. Oceanog.*, **6**, 781–800.

Paltridge, E. W. (1980). Cloud-radiation feedback to climate. *Quart. J. Roy. Meteor. Soc.*, **106**, 895–899.

Parkinson, C. L., and W. W. Kellogg (1979). Arctic sea ice decay simulated for a CO_2-induced temperature rise. *Climatic Change*, **2**, 149–162.

Parkinson, C. L., and W. M. Washington (1979). A large-scale numerical model of sea ice. *J. Geophys. Res.*, **84**, 311–337.

Pimentel, D. (1980). Increased CO_2 effects on the environment and in turn on agriculture and forestry. In, AAAS (ed.), Workshop on environmental and societal consequences of a possible CO_2-induced climate change. CONF-7904143, US Department of Energy, Washington, DC.

Pimentel, D., and M. Pimentel (1978). Dimensions of the world food problem and losses to pests. In, D. Pimentel (ed.), *World Food, Pest Losses, and the Environment, D. Pimentel.* Westview Press, Boulder, Colorado, USA.

Pittock, A. B., and M. J. Salinger (1981). Towards regional scenarios for a CO_2-warmed earth. *Climatic Change*, **4**, 23–40.

Potter, G. L. (1979). Zonal model calculation of the climatic effect of increased CO_2. Paper presented at Symposium on Environmental and Climatic Impact of Coal Utilization, Williamsburg, Virginia.

Ramanathan, V. (1981). The role of ocean–atmosphere interactions in the CO_2 climate problem. *J. Atmos. Sci.*, **38**, 918–930.

Ramanathan, V., and J. A. Coakley (1978). Climate modeling through radiative–convective models. *Rev. Geophys. Space Phys.*, **6**, 465–489.

Ramanathan, V., M. S. Lain, and R. D. Cess (1979). Increased atmospheric CO_2: zonal and seasonal estimates of the effect on the radiation balance and surface temperature. *J. Geophys. Res.*, **84**, 4949–4958.

Ramirez, J. M., C. M. Sakamoto, and R. E. Jensen (1975). Wheat. In, *Impacts of Climatic Change on the Biosphere*, CIAP Monograph 5, Part 2. *Climatic Impact Assessment Program.* US Department of Transportation, Washington, DC.

Revelle, R., and W. Munk (1977). The carbon dioxide cycle and the biosphere. In, *Energy and Climate.* National Academy of Sciences, Washington, DC.

Rotty, R. M. (1973). Commentary on and the extension of calculative procedure for CO_2 production. *Tellus*, **25**, 508–517.

Rotty, R. M. (1971). Global carbon dioxide production from fossil fuels and cement, A.D. 1950; A.D. 2000. In, N. R. Anderson and A. Maldhoff (eds.), *The Fate of Fossil Fuel CO_2 in the Oceans.* Plenum Press, New York.

Rotty, R. M. (1978). The atmospheric CO_2 consequences of heavy dependence on coal. In J. Williams (ed.), *Carbon Dioxide, Climate and Society.* Pergamon Press, Oxford, England.

Rotty, R. M. (1979a). Present and future production of CO_2 from fossil fuels—a global appraisal. In, Workshop on the Global Effects of Carbon Dioxide from Fossil Fuels. CONF-770385. US Dept. of Energy, Washington, DC.

Rotty, R. M. (1979b). CO_2 emissions from synfuel and oil from shale. Unpublished notes. Institute for Energy Analysis, Tennessee.

Rotty, R. M. (1982). Distribution of and changes in industrial carbon dioxide production. *J. Geophys. Res.* (in press).

Rotty, R. M., and G. Marland (1980). Constraints on fossil fuel use. In, W. Bach, J. Pankrath, and J. Williams (eds.), *Interactions of Energy and Climate.* Reidel, Dordrecht, Holland.

110

Schneider, S. H. (1975). On the carbon-dioxide–climate confusion. *J. Atmos. Sci.*, **32**, 2060–2066.

Schneider, S. H. (1979). So what if the climate changes? Paper presented at AAAS Annual Meeting, Houston.

Schneider, S. H. (1980). Comparative rise assessment of energy systems. In, W. Bach *et al.*, *Renewable Energy Prospects*. Pergamon Press, Oxford, England.

Schneider, S. H., and R. S. Chen (1980). Carbon dioxide warming and coastline flooding: Physical factors and climatic impact. *Ann. Rev. Energy*, **5**, 107–140.

Schneider, S. H., and S. L. Thompson (1981). Atmospheric CO_2 and climate: Importance of the transient response. *J. Geophys. Res.*, **86**, 3135–3147.

Siegenthaler, U., and H. Oeschger (1978). Predicting future atmospheric CO_2 levels. *Science*, **199**, 388–395.

Singh, J. J. and A. Deepak (eds.) (1980). *Environmental and Climatic Impact of Coal Utilization*. Academic Press, New York.

Slade, D. H. (1979). Summary of the carbon dioxide effect research and assessment program. US Department of Energy, Washington, DC.

Smagorinsky, J. (1977). Modeling and predictability. In, *Energy and Climate*, National Academy of Sciences, Washington, DC.

SMIC (1971). Inadvertent climate modification. Report of the study of man's impact on climate. MIT Press, Cambridge, Massachusetts.

Smith, I. M. (1982). Carbon dioxide—emissions and effects. Report No. ICTIS/TR 18, IEA Coal Research, London.

Steinberg, M., A. S. Albanese, and V.-D. Dang (1978). Environmental control technology for carbon dioxide. Paper presented at 71st Annual Meeting of American Institute of Chemical Engineers, Miami, Florida.

Steinberg, M., and A. S. Albanese (1980). Environmental control technology for atmospheric carbon dioxide. In, W. Bach, J. Pankrath and J. Williams (eds.), *Interactions of Energy and Climate*. D. Reidel Publ. Co., Dordrecht, Holland.

Takahashi, K., and M. Yoshino (eds.) (1978). *Climate Change and Food Production*. University of Tokyo Press, Japan.

Thompson, L. M. (1975). Weather variability, climatic change and grain production. *Science*, **188**, 535–541.

Thompson, S. L., and S. H. Schneider (1979). A seasonal zonal energy balance climate model with an interactive lower layer. *J. Geophys. Res.*, **84**, 2401–2414.

USDOE, 1979: Summary of the carbon dioxide effects research and assessment program. U.S. Department of Energy, Washington, DC.

Washington, W. M., and V. Ramanathan (1980). Climate response due to increased CO_2: Status of model experiments and the possible role of the oceans. Carbon Dioxide and Climate Research Program's Progress and Planning Meeting, April 1980, US Department of Energy, Washington, DC.

Washington, W. M., A. J. Semtner, G. A. Meehl, D. J. Knight, and T. A. Mayer (1980). A general circulation experiment with a coupled atmosphere, ocean and sea ice model. (Submitted to *J. Phys. Oceanogr.*)

WCC (1970). *Proceedings of the World Climate Conference*. WMO Publication No. 537, World Meteorological Organization, Geneva.

Wetherald, R. T., and S. Manabe (1980). Cloud cover and climate sensitivity. *J. Atmos. Sci.*, **37**, 1485–1510.

Whittaker, R. H., and G. E. Likens (1973). Carbon in the biota. In, G. M. Woodwell and E. V. Pecan (eds.), *Carbon and the Biosphere*. CONF-720510, US Atomic Energy Commission, NTIS, Springfield, Virginia.

Wigley, T. M. L., and P. D. Jones (1981). Detecting CO_2-induced climatic change. *Nature*, **292**, 205–208.

Wigley, T. M. L., P. D. Jones, and P. M. Kelly (1980). Scenario for a warm high-CO_2 world. *Nature*, **283**, 17–21.

Williams, J. (ed.) (1978). *Carbon Dioxide, Climate and Society*. Pergamon Press, Oxford, England.

Williams, J. (1980). Anomalies in temperature and rainfall during warm Arctic seasons as a guide to the formulation of climate scenarios. *Climatic Change*, 2, 249–266.

Wiser, H. L. (1978). Scientific information required by policymakers and regulating decisionmakers. In, J. Williams (ed.), *Carbon Dioxide, Climate and Society*. Pergamon Press, Oxford, England.

Woodwell, G. M. (1978). The global carbon dioxide question. *Sci. Amer.*, **238**, 34–43.

Woodwell, G. M. (ed.) (1980). Measurement of changes in terrestrial carbon using remote sensing. CONF-7905176 UC-11, United States Department of Energy, Washington, DC.

Woodwell, G. M., R. H. Whittaker, W. A. Reiners, G. E. Likens, C. C. Delwiche, and C. B. Botkin (1978). The biota and the world carbon budget. *Science*, **199**, 141–146.

World Climate Programme (1981). On the assessment of the role of CO_2 on climate variations and their impact. Report of Joint WMO/ICSU/UNEP Meeting of Experts, World Meteorological Organization, Geneva.

Zimen, K. E. (1979). The carbon cycle, the missing sink, and future CO_2 levels in the atmosphere. In, W. Bach *et al.* (eds.), *Man's Impact on Climate*. Elsevier, Amsterdam.

CHAPTER 4

The Effect of Waste Heat Release on Climate

4.1 INTRODUCTION

Energy conversion, whether it involves converting coal to electricity at a power plant or electricity to light in a home, releases heat to the environment. In the case of power plants, conversion efficiencies are much less than 100%. In efficient fossil-fuel-fired steam-electric plants only about 40% of the heat produced by combustion is converted to electricity; about 45% of the heat is discharged to cooling water and the remaining 15% is lost in the plant, up the stack and in the ash. Nuclear power plants of the type now in use or planned have efficiencies of about 33%, about 62% of the heat produced is discharged into cooling water and the remaining 5% is lost in the plant. At the final point of use of energy, whether it is in the form of coal, oil, gas, or electricity, almost 100% of the energy is released as heat.

This chapter considers several aspects of the effect of such heat releases on the climate. Firstly, it compares, on a globally averaged scale, the heat released by human activities with the magnitude of other energy sources. Next, it discusses effects of power plants on the local meteorological conditions. This is followed by an evaluation of the potential effects of proposed 'power parks', groupings of power plants. The effects of heat releases from present urban-industrial areas are then briefly described. Finally the possible consequences of large waste heat released on the global climate are discussed.

4.2 THE MAGNITUDE OF HEAT RELEASED BY HUMAN ACTIVITIES

When averaged over the globe, the amount of heat released by human activities is a small fraction of the solar radiation absorbed at the earth's surface. Figure 4.1 from Schneider and Dennett (1975) shows the balance between incoming solar radiation and outgoing terrestrial infrared radiation and the distribution of energy in the global system. All of the values are percentages of the incoming solar radiation. The global average solar incoming radiation is about 350 W m^{-2} and the global average solar energy absorbed by the earth's surface is about 160 W m^{-2}.

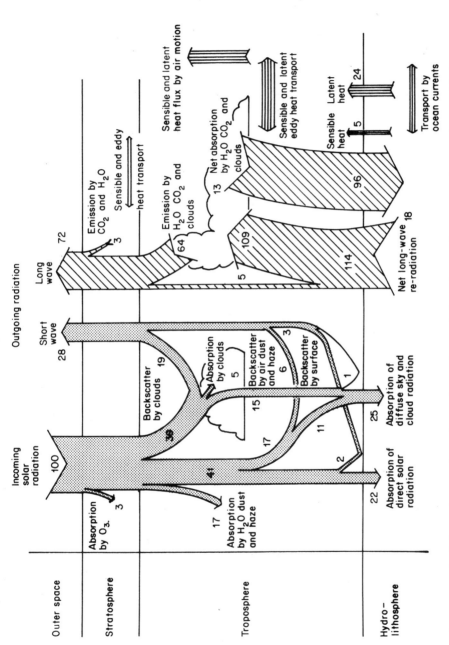

Figure 4.1   The balance between incoming and outgoing radiation and the distribution of energy in the global system.
Source: Schneider and Dennett (1975)

The following values, taken from Munn and Machta (1979) compare the heat releases of human activities with other energy sources.

Global average solar radiation at the outer edge of the atmosphere		350 W m^{-2}
Global average solar radiation absorbed at the earth's surface		160 W m^{-2}
1970 energy use distributed evenly over the globe		0.016 W m^{-2}
1970 energy use distributed evenly over the continents		0.054 W m^{-2}
Annual global energy flow from the earth's interior		0.06 W m^{-2}
Heat from major USA cities	summer	20 to 40 W m^{-2}
	winter	70 to 210 W m^{-1}

The above figures show that on a global average the use of energy was 10^{-4} of the solar energy absorbed at the earth's surface. However, at individual places on the earth's surface the heat release due to human activities is of the same order of magnitude as or larger than the absorbed solar energy. This fact is illustrated by the figures for heat release in large cities; it will be discussed further in Section 4.5.

Returning to the global average figures, we note that while heat releases appear to be negligible on a global scale at the present time, large releases in the future could become more significant. Schneider and Dennett (1975) and Kellogg (1977) have considered the heat releases from a global population of 20,000 million with a *per capita* consumption of 20 kW. In this case the global average heat release would be about 0.5% of the global average solar radiation absorbed at the earth's surface. Schneider and Dennett (1975) have suggested, on the basis of a simple calculation, that this would give rise to a global average surface temperature increase of 0.4 K. Kellogg (1977) refers to a number of climate model results (e.g., Budyko, 1969; Sellers, 1969; Wetherald and Manabe, 1975) which suggest that a 1% increase in the heat available to the system would result in about 2 K increase (with a factor of 2 uncertainty) in the mean surface temperature. Clearly, the assumptions of a population of 20,000 million and a *per capita* energy consumption of 20 kW yr/yr make the calculations only of academic interest, since these values are unlikely to be reached. Such assumptions were made at a time when it was believed that the world might need such large amounts of energy. Now they only serve to show the absolute and unrealistic upper boundary.

Penner (1976) used the global heat-balance equation of Budyko (1969) to show that heat addition associated with global energy use in the year 2050 would cause a mean global temperature rise of 0.27 K, assuming a 20 kW *per capita* energy use and a world population of 10,000 million.

4.3 THE EFFECT OF POWER PLANTS

As mentioned above, power plant efficiencies are of the order of 30–40% and most of the remaining heat that is generated must be removed from the power plant by a cooling system. Several types of cooling systems are used; these have been discussed by Bach (1979), Bhumralkar and Williams (1980), and others. The simplest type of cooling system is once-through cooling, in which water is drawn from a natural water body, used to cool the power plant, and reinjected into the water body (river, pond, lake). The disadvantages of this type of system include fogging and icing in some locations. Since the availability of water bodies for such cooling systems is limited, more power stations in the future will rely on cooling towers; thus, their effect on climate will be given more emphasis here.

One must distinguish between wet and dry cooling towers. In wet cooling towers, the heat is released into the atmosphere in both latent and sensible forms. Of the two subtypes, natural-draught towers generally produce less fogging or icing. Mechanical-draught towers, which are typically one-fifth as high as the natural-draught towers, increase the relative humidity in the vicinity. In dry cooling towers sensible heat is released into the atmosphere. It is claimed that this type of cooling system produces less visible clouds, but Bach (1979) cites results of numerical simulations suggesting that dry cooling towers give rise to more anomalous convective cloud formation than wet cooling towers.

Koenig and Bhumralkar (1974) have listed the potential problems associated with cooling towers as:

- restriction of sunlight by a visible plume;
- deposition of detrimental chemicals in cooling waters on to surrounding areas;
- restriction of visibility by visible plumes reaching the ground (fogging);
- change in the amount and distribution (spatial and temporal) of precipitation;
- initiation of severe weather, such as tornadoes or thunderstorms.

Reviews of the observed atmospheric effects of power plants have been given, for example, by Hanna (1978) and Bhumralkar and Williams (1980). The most frequently observed change is due to low stratus or fog formation in the plumes from cooling ponds and towers. Hanna (1974) found that ground fog occurs within 200 m of the Oak Ridge mechanical-draught towers 40% of the time and that a stratus cloud extends several kilometres downwind on naturally rainy or foggy days. Drift deposition is not believed to be a problem in view of the development and use of efficient drift eliminators in the cooling towers, although this may not be true when sea water is used as the cooling agent.

According to Hanna (1978), only a few reports of precipitation enhancement in the vicinity of power plants have been published. Kramer et al. (1976) report that, during the winter of 1975–1976, snowfall was observed from plumes of large natural-draught cooling towers. Similarly, snowfalls downwind of cooling towers have been reported by Culkowski (1962) and Ott (1976). On the other hand, Martin (1974) reports a study of local weather records near a 2 GW(e) power station with eight natural-draught cooling towers. For the four years of operation of the power station, he concluded that emissions had not affected the values of total rainfall, hours of bright sunshine, or incidence of morning fog recorded by stations at distances of 4 km or more from the power plant. Landsberg (1977) suggests that the difference in precipitation due to emissions from present-day power plants is small enough that it can be found to be significant only during special case studies with short time and space scales.

Weber (1978) describes a study of the effect of a nuclear power plant with a once-through cooling system and one with mechanical-draught cooling towers on the average air temperature in the vicinity of the plants. The plants produce 1,100 MW(e) and 700 MW(e), respectively. He concluded that, on the basis of the data available and the analyses made so far, there has been no detectable average increase in temperature at meteorological stations near either plant.

A further potential effect of power plants is the concentration of vorticity by large buoyant plumes and observations of vortices have been reported (Hanna, 1978). Small waterspouts were observed occasionally over one cooling pond (Everett and Zerbe, 1977) and vortices have also been observed, for instance, over many large wildfires (e.g., Taylor et al., 1973). Hanna (1978) points out that, as the output of power production facilities increases beyond 10,000 MW, the incidence of associated whirlwinds can be expected to increase.

In addition to the observations of the atmospheric effects of power plants, numerical models have been developed to explain and possibly predict such effects. They have been reviewed by Hanna (1978) and Bhumralkar and Williams (1980). On the small scale, there are models of cooling-tower plumes and drift deposition, which, according to Hanna (1978), can estimate plume length and plume rise within a factor of two and drift deposition within an order of magnitude. Reviews indicate that about 10 of each of these two types of model are satisfactory.

For larger scales, 1 to 10 km, cloud-growth models can be used to estimate cooling-tower plumes. One- and two-dimensional models have been developed and used to describe plumes. Other models exist for estimating the occurrence of fog downwind of cooling ponds and towers.

Predictions of the development of whirlwinds or vortices at power parks can be made using a similarity criterion developed by Briggs (1974). It predicts that existing single cooling towers would have no whirlwinds, while a cluster of 20 cooling towers would cause them to form. The criterion determines whether a given buoyant source with access to large-scale natural vorticity will concentrate the vorticity into a whirlwind.

Hanna and Gifford (1975) summarize the effects of power plants of local meteorological conditions by pointing out that the maximum amount of electrical power currently generated at a single power station is about 3 GW(e) and the atmospheric effects of current heat dissipation rates there are not serious problems, especially beyond the scale of the power station, provided that efforts are made to design the facility such that downwash is eliminated, drift is minimized, and plume rise is maximized. They find that fog formation and drift deposition are generally localized and that, although clouds are observed, no significant changes in rainfall in areas of study have been detected. However, concern has been expressed over the possible climate effect of 'power parks', releasing much more heat than present power plants; this problem is discussed in the next section.

4.4 THE EFFECT OF POWER PARKS ON LOCAL AND REGIONAL WEATHER AND CLIMATE

In contrast to the present situation of dispersed sites for power generation, some proposals have been made for future concentrated sites, here referred to as power parks. For example, Burwell *et al.* (1979) have proposed a nuclear-power-plant siting policy with a few large concentrated sites, which would to some extent isolate nuclear energy activities, while augmenting the strength of the institutions responsible for managing them. They also feel that such sites would have an element of permanence, opening new options for managing low-level wastes and reactor decommissioning. They suggest that some of the sites might produce as much as 20 GW(e).

Several recent studies have addressed the effect of proposed 10–50 GW(e) power parks on climate (Rotty, 1974; Rotty *et al.*, 1976; Hanna and Gifford, 1975; Koenig and Bhumralkar, 1974; Bhumralkar and Alich, 1976; Hosler and Landsberg, 1977). The effects of such large releases of heat can be assessed very crudely by analogy with effects of comparable sources of heat and moisture, such as islands heated by solar radiation, urban-industrial complexes, forest fires, and phenomena such as volcanoes. In addition, the effects have been studied by applying numerical models.

Table 4.1, based on figures in Hosler and Landsberg (1977), compares estimates of man-made and natural energy releases. It appears that a 20 GW(e) power park (releasing 40 GW heat locally) might produce atmospheric effects of the same magnitude as large cities or the island of Aruba. The energy release

Table 4.1   Comparison of man-made and natural energy releases. Source: Hosler and Landsberg (1977)

Heat source	Area (km^2)	Approximate MW equivalent input to the atmosphere
St Louis	250	16,100
Chicago	1,800	52,700
Aruba	180	35,000
One-megaton nuclear device (heat dissipated over 1 hour)		1,000,000
20,000 MW power park	26	40,000
Cyclone (1 cm of rain/day)	10×10^5	100,000,000
Tornado (kinetic energy)	10×10^{-4}	100
Thunderstorm (1 cm of rain/30 min)	100	100,000
Great Lakes snow squall (4 cm of snow/hour)	10,000	10,000,000

from such a power park is of the same order of magnitude as some mesoscale atmospheric phenomena—tornadoes and thunderstorms. It is unlikely that a 20 GW(e) power plant could have an effect on cyclone-scale atmospheric conditions, but it could cause local/regional changes in cloudiness and precipitation, as observed, for example, in the case of St Louis (Changnon, 1973; and see next section) or the island of Aruba. The figures in Table 4.1 show that the total energy content of a tornado is not larger than that of a power park. As pointed out in the last section, tornado formation requires vorticity production; for heat releases from large power parks, such vortex formation is possible, in contrast to the effects of single cooling towers.

Church *et al.* (1980) have made observations of the atmospheric effects of a total heat output of 1 GW produced by an array of 105 oil burners referred to as the Météotron. They observed three types of vortices: large counter-rotating rolls in the downstream plume; intense small-scale vortices resembling very strong dust devils; and very large columnar vortices. They conclude that generating vortices of moderate intensity is to be expected in large plumes from forest fires or industrial operations.

A different comparison of man-made and natural energy releases is given in Figure 4.2, from Fortak (1979). Here the area of a heat source is plotted on the abscissa and the heat release is plotted on the ordinate. We see that power parks (here referred to as energy centres, EC) with a heat emission of 10^4–10^5 MW and radius of order 0.5 to 3 km would produce vertical air velocities between those over forest fires and volcanoes or of cumulus and cumulonimbus clouds. Rotty (1974) has cited observations showing development of whirlwinds, tornadoes, thunderstorms, and cumulus-type clouds over heat sources such as volcanoes and intense fires. Fortak (1979) has concluded on the basis of this figure that clustered power parks are undesirable from a meteorological point of view. The shaded area in the figure shows the range within which the cooling-tower spacing for power parks should be planned. The minimum spacing for a cooling-tower grid is about 5 km (ΔE).

Koenig and Bhumralkar (1974) used a model of local atmospheric conditions to investigate the effect of a 36 GW(e) power park proposed for Louisiana. They found that the heat release caused changes in the model temperature and moisture fields (3–4 °C rise over 6 km, 1–2 g/kg increase of water-vapour mixing ratio) large enough to initiate convection (and therefore cloudiness and rainfall). They also concluded that the downwind modification of the cloudiness and rainfall would be more consistent and visible for an energy park than for an urban heat island. Bhumralkar and Alich (1976) have also modelled the effects of waste heat from a proposed 36 GW(e) power park and found that significant weather modification could occur. An updated version of the same model has been used by Bhumralkar *et al.* (1979) to investigate the effect of a large (1,000 km^2) solar energy conversion park; the results are described in Chapter 6.

Hosler and Landsberg (1977) describe some model results of Deaven showing the effects of a 20 GW(e) power park. The heat input caused significant perturbations in isentropic surfaces and streamlines. They suggest that a three-dimensional

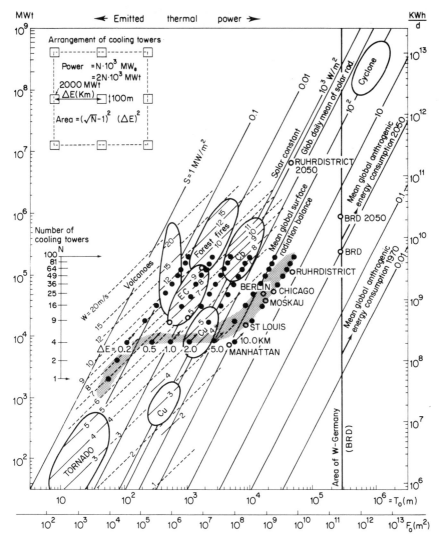

Figure 4.2   Energy flux densities of natural and anthropogenic phenomena. Source:
Fortak (1979)

perturbation of the magnitude found in Deaven's two-dimensional model would
form a vortex flow and that inclusion of moisture effects would result in additional
instability.

Hanna and Gifford (1975) made an extensive survey of the potential effects of
power parks. Their comparison with the energy production of natural
atmospheric processes, similar to that shown in Table 4.1, showed that the heat
releases at 10–50 GW(e) power parks lie between thunderstorms and squall lines
on the energy intensity scale. Comparison with the effects of other heat sources,

such as wildfires, oil fires, natural convection, and volcanoes, and results of computations suggest that power parks could concentrate vorticity. Hanna and Gifford also find that plume rise from a group of cooling towers will be at least 10% greater than the rise from a single tower if the towers are spaced closer than 300 m. For a power park location in the southeast US, they predict that a condensed plume would occur about 20% of the time at distances of 100 km from natural-draught towers dissipating about 80 MW of heat. In addition, they estimate that excess ground fog would occur about 2% of the time at a given location within 100 km of the power park and that naturally occurring precipitation would be augmented.

In summary, the results of model and analogue studies and comparisons with natural phenomena suggest that the principal effects of the release of large amounts of waste heat from power parks would be, on the local and perhaps on the regional scale, significant changes in cloudiness and precipitation with a possible increase in the probability of severe weather.

4.5   EFFECTS OF URBAN-INDUSTRIAL AREAS

Probably the best known examples of the results of man's activities on the atmosphere are the observed effects in urban areas, in particular the 'urban heat island'; that is, the tendency for an urban area to be warmer than the surrounding rural area, especially during early evening hours, when the wind is light. Most recently this effect has been observed using satellite imagery. Matson *et al.* (1978) show that on 28 July 1977 an unusually cloud-free night-time thermal infrared image of the midwestern and northeastern United States enabled detection of more than 50 urban heat islands. Analysis of digital data from the satellite for selected cities showed the maximum urban–rural temperature differences listed in Table 4.2.

Oke (1982) has discussed the energetic basis of the urban heat island in detail. He shows that the urban heat island is a thermal anomaly with horizontal, vertical, and temporal dimensions. It has been observed in virtually all settlements, large and small, where it has been sought. Oke emphasizes that in the middle latitudes, where it has been most studied, its characteristics are related to the intrinsic nature of the city (e.g., size, building density, and land-use distribution) and to external influences (e.g., the climate). Climate alterations by urbanization have also been discussed recently by Landsberg (1979, 1981) and Oke (1980).

As Landsberg (1975), for example, has pointed out, the urban heat island is a result of many factors. The city landscape causes an increase in surface roughness in comparison with rural areas and this leads to changes in the vertical wind profile, the wind speeds near the surface being reduced. The spongy, often moist, soil cover of rural areas, which has low heat conductivity, is converted into an impermeable surface layer, with a high capacity for conducting heat and, because of generally low reflectivity, high absorptivity for radiation. The changes in surface conditions also lead to more rapid run-off of precipitation and therefore to a reduction in local evaporation, which is equivalent to a further heat gain. In addition, heat release from energy use contributes to the heat island.

Table 4.2   Maximum heat-island effect of 11 metropolitan areas.
Source: Matson *et al.* (1978)

City	Maximum heat-island effect (urban–rural temperature) (°C)	Population[a] (×10⁶)
Louisville, KY (SDF)	6.5	0.89
Baltimore, MD (BAL)	5.2	2.14
Washington, DC (DCA)	5.2	3.02
Cincinnati, OH (CVG)	5.1	1.38
Indianapolis, IN (IND)	4.5	1.14
Dayton, OH (DAY)	4.5	0.84
St Louis, MO (STL)	4.4	2.37
Richmond, VA (RIC)	3.8	0.57
Columbus, OH (CMH)	3.3	1.07
Kansas City, MO (MCI)	3.2	1.30
Petersburg, VA —	2.6	0.04

a. 1974 metropolitan populations based on US Census Standard Metropolitan Statistical Areas (SMSA) as defined in the 1970 census.

In view of the theme of this report, it is clearly important to discuss the role of heat release in forming urban climates. However, studies of the atmospheric effects of the urban heat island, for example on downstream cloudiness and precipitation, may also be important as analogies for the effects of future power parks. Similarly, modelling urban heat islands lead to developing methods applicable to other heat-release studies. Since only a brief description can be given here, it should be pointed out that more extensive reviews of urban climate characteristics and models have been given, for instance, by Oke (1974), Landsberg (1975), Garstang *et al.* (1975), and Hosler and Landsberg (1977).

In some high-latitude, or densely populated, or industrialized cities, the anthropogenic heat release can be higher than the net radiation balance, especially in the winter season. Table 4.3 shows this by comparing the energy consumption density with the average net radiation in several American and European urban-industrial areas. We see that, on a local scale (up to 10^3 km^2), the artificial energy flux density has the same order of magnitude as the natural net radiation, and in very heavily industrialized cities, such as Moscow, it is larger. When the values are examined on a seasonal basis, it is seen that in the cold season the anthropogenic heat release may well dominate as the energy source, while in summer it may be insignificant in comparison with the net radiation balance (Oke, 1974). Computations based on theoretical models and empirical data confirm that during the winter in high latitudes a major part of the heat island can be accounted for on the basis of anthropogenic heat releases (e.g., Leahey and Friend, 1971; Oke and East, 1971).

Certain locations may have anomalously high anthropogenic heat releases. For example, Oke and Hannell (1970) investigated the urban heat island in Hamilton,

Table 4.3   Energy consumption density (ECD) and average net radiation for certain urban-industrial areas. Source: SMIC (1971)

Urban-industrial area	Area (km²)	ECD (W/m²)	Average net radiation (W/m²)
Nordrhein-Westphalen	34,039	4.2	50
Same, industrial area only	10,296	10.2	51
West Berlin	234[a]	21.3	57
Moscow	878	127.0	42
Sheffield (1952)	48[a]	19.0	46
Hamburg	747	12.6	55
Cincinnati	200[a]	26.0	99
Los Angeles County	10,000	7.5	108
Los Angeles	3,500[a]	21.0	108
New York, Manhattan	59	117.0	93
Fairbanks, Alaska	37	18.5	18

a. Building area only.

Canada. They found a large cell of warmer air centred over a heavily industrialized area, because of the release of heat in steel making. The largest steel works estimated that its heat release was about 2.5 GW. The sensible heat flux was estimated to be between 370 and 560 W m⁻² and the latent heat flux about 46 W m⁻², in comparison with the annual net radiation at Toronto of 62 W m⁻². The sensible heat flux was larger than the net radiation balance on a summer's day.

Figure 4.3 shows an idealized description of the urban heat island at night with clear calm weather. There is increased convection over the urban area, which is

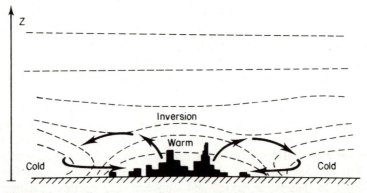

Figure 4.3   Idealized pattern of the atmospheric circulation above a city on a clear, calm night. The diagram shows the urban heat island and radiative ground inversions in rural areas. The dashed lines represent isotherms and the arrows show the direction of the air circulation. Source: Landsberg (1975)

also especially observable during the daytime. The updraft, together with the additional water vapour released into the urban atmosphere by combustion and power-plant cooling systems leads to increased cloudiness over the city (Landsberg, 1975). It is also an important factor in the increased rainfall reported over cities.

In addition to the observed temperature difference between the urban area and the surrounding rural area, there are observable effects on other climatic variables. For example, Cologne has an average of 34% fewer days with temperature minima below 0 °C than the surrounding area; in London, Kew has an average of 72 more days with frost-free screen temperatures than rural Wisley (Barry and Chorley, 1968). This also means that less snow falls (since it falls as rain or sleet) in urban areas and it melts away faster. Generally, it is claimed that urban wind speeds are lower, but this depends on the prevailing wind characteristics and the structure of the urban surface. In addition, at night greater mechanical turbulence over the city can mean that the higher wind speeds aloft are transferred to the lower-level air by turbulent mixing.

According to Oke (1980), the effect of cities upon cloud and precipitation is a controversial topic. The present consensus is that there is reasonable statistical and other support for the hypothesis of precipitation enhancement by large cities, but that proof, including knowledge of the governing physical mechanisms, remains to be exhibited. There is considerable evidence that enhancement does occur downwind of urban areas and convective events are most susceptible to modification (Oke, 1980). The typical size of the increase in precipitation is about 5–30% on an annual basis and the maximum effect is usually located a few tens of kilometres downwind of the urban area.

The effects of urban areas on cloudiness and precipitation have been reviewed in detail by Garstang et al. (1975). Over the city, combustion sources add large amounts of condensation and freezing nuclei to the atmosphere, which increase the probability of fog occurrence, as found, for example, for Prague by Geiger (1965) and for London by Chandler (1965). Many investigations have confirmed an urban-induced precipitation increase. Schmaus (1927) found an 11% increase in both light rainfall and the number of convective showers over Munich in comparison with the surrounding rural area. Several authors have reported weekly cycles in precipitation, with drier weekends over urban-industrial areas (Garstang et al., 1975). A large anomaly was reported by Changnon (1969) for La Porte, about 35 km downwind from Chicago, Illinois, and Gary, Indiana, with a 31% increase of annual precipitation. Changnon (1980) has reviewed the evidence for the La Porte precipitation anomaly. He finds that local records suggest an increase in warm-season rainfall, thunderstorms, and hail during the 1935–1965 period. A variety of recent studies suggest that the La Porte anomaly began to shift locale in the 1950s and then disappeared in the 1960s. Changnon concludes that the anomaly was due to urban influences on the atmosphere, but the anomaly either ended or shifted into Lake Michigan (where it cannot now be detected) as the general circulation pattern changed, leading to fewer cyclonic passages since 1960. Atkinson (1968) has shown that the higher incidence of thundershowers over London can be attributed to urban effects alone. Table 4.4 shows the results of a study of the effects of nine US cities on summer rainfall (Changnon, 1973). At six of the cities an increase (9–27%) in summer rainfall is observed, usually downwind of the urban centre. At the same six cities a 10–47% increase in the number of days with thunder was also observed.

Garstang et al. (1975) have pointed out that both urban areas and ocean islands have a heat-island effect, because of their thermal and physical properties. The atmospheric effects of an urban heat island are in fact more difficult to study because of the complexities introduced by anthropogenic heat release, building configuration and geometry, street orientation, pollution, etc., in urban areas. Models have, however, been developed for both types of heat island. Garstang et al. (1975) distinguish between four broad categories of heat-island model: statistical, energy-balance, turbulent-mixing, and dynamic. A detailed review of the methods and results is given by Garstang et al. and Oke (1974) has discussed the history and state of the art of heat-island and urban climate modelling.

Oke (1980) has discussed the effects of urban areas on larger-scale climates. He points out that the total area of the earth that is urbanized amounts to 0.2% of the total earth area. The present rate of change is estimated to be about 2×10^4 km^2 yr^{-1} (4×10^{-3}% yr^{-1}). Thus, Oke concludes that the surface changes accompanying urbanization do not appear to have any significant effect on global-scale climate. However, he points out that this should not be construed to suggest that cities do not affect climates on scales beyond the local scale. Emission of pollutants (gaseous, aerosol, and thermal) from urban areas has effects on all climatic scales. Also, a number of effects extend downwind of urban areas up to 100 km. Oke also suggests that urban areas may modify the path, speed, and other characteristics of mesoscale and synoptic systems passing across or near them. Lastly, he suggests that the clustering of conurbations gives rise to the

Table 4.4 Summary of urban effects on summer rainfall in nine cities. Source: Changnon (1973)

City	Effect observed	Maximum change (mm)	Maximum change (%)	Approximate location
St Louis	Increase	41	15	16–19 km downwind
Chicago	Increase	51	17	48–56 km downwind
Cleveland	Increase	64	27	32–40 km downwind
Indianapolis[a]	Indeterminate	—	—	—
Washington	Increase	28	9	48–64 km downwind
Houston[b]	Increase	18	9	Near city centre
New Orleans[b]	Increase	46	10	NE side of city
Tulsa	None	—	—	—
Detroit	Increase	20	25	City centre

a. Sampling density not adequate for reliable evaluation.
b. Urban effect identified only with air mass storms—apparently little or no effect in frontal storms.

probability that the effects of one city will overlap those of one or more others and produce a cumulative impact.

4.6   THE EFFECT ON GLOBAL CLIMATE OF WASTE HEAT RELEASE FROM POWER PARKS

Earlier sections have discussed the effect on local and regional weather and climate of waste heat release from power parks. In contrast, the next two sections will examine the possible effect on global climate of large releases of waste heat. The global considerations presented in Section 4.2 suggest that global energy use would have to be extremely large for heat release to cause significant changes in the global average surface temperature. However, such global considerations are misleading for two reasons. Firstly, it has been pointed out in Chapter 2 that global changes are the sum of regional changes, some of which will be larger than the global average change and some of which will be of opposite sign. Secondly, the major forcing mechanism in the climate system is differential heating. It is clear that heat release from mankind's activities is not and will not be evenly distributed over the earth's surface. The heat release is and, with high probability, will be concentrated in certain areas (power parks, urban-industrial complexes, megalopolitan areas, etc.). Thus, the possible effect of more localized heat release on the atmospheric or oceanic circulation must also be examined. Basically this amounts to considering whether large-scale effects could occur if heat releases in particular places were large enough to interfere with weather systems on the scale of cyclones and anticyclones and cause further anomalies downstream in areas where heat release did not occur. Sawyer (1965) has pointed out that anomalous heat sources could be considered as plausible causes of long-term climate anomalies if they are: (a) of large extent; (b) persistent over periods comparable with those of weather systems; (c) of sufficient magnitude. This suggests that the area of heating should be comparable with the scale of the temperature and pressure anomalies that they produce, say 1,000 km or more across. They should persist for periods of the order of a month at least. As a rough measure of the size of a significant heat release, Sawyer suggested that they should release an amount equivalent to a significant fraction (at least one-tenth) of the normal outgoing long-wave radiation to space, about 200 W/m^2.

It is already observed that certain natural anomalies meet these criteria and thus have an influence on the climate system. In particular, large-scale anomalies in ocean surface temperature have been seen to have climatic effects; some of the relevant studies have been described in Chapter 2.

The rest of this chapter considers the potential effect of man-made anomalies. Such effects have been evaluated by using models of atmospheric circulation. As discussed in Chapter 2, the model studies involve comparing the simulated atmospheric circulation for a model integration without any added perturbation (a 'control case') with an integration involving a perturbation (e.g., waste heat). Because of model shortcomings, one cannot predict how the atmosphere would respond to a particular heat input, but the model studies are a guide to the order of magnitude of any response.

The first studies of the effect of waste heat release on the simulated atmospheric circulation were made by Washington (1971), using the NCAR general circulation model (GCM). He investigated the response of the model atmosphere to an addition of 24 W m^{-2} over all continental and ice regions and found a 1–2 °C increase in average surface temperature with an 8 °C increase over Siberia and northern Canada. However, the total heat release was very unrealistic, both in magnitude and distribution.

Washington (1972) used a more realistic input of energy. He assumed a *per capita* energy use of 15 kW, a population of 20,000 million, and a heat-input distribution according to present population density. Four simulations were carried out with the model: a control case; an integration with the stipulated heat addition; an integration with this heat amount but with a minus sign; and a case the same as the control case but with a small random change in the initial conditions. He concluded that the differences between the simulated atmospheric conditions in the control case and the case with the positive waste-heat addition were not larger than the differences between the cases with positive and negative heat addition, or the control case and randomly perturbed control case. That is, the signal (response to waste-heat addition) was found to be not larger than the noise (inherent variability) of the model.

Using the GCM developed at the UK Meteorological Office, a number of scenarios of waste-heat release have been investigated (Murphy *et al.*, 1976; Williams *et al.*, 1977a and b, 1979). For these simulations the waste-heat release was concentrated at groups of four model grid points in ocean areas; these localized points of heat release are referred to as energy parks or islands. Figure 4.4 shows the locations of the three energy parks that were studied. Table 4.5 lists the amount of heat input and the combination of energy parks used in each perturbed integration. In addition to the five perturbed cases, three control cases of the same model were available for comparison.

A completely realistic simulation of the effects of such energy parks would require using a linked atmosphere–ocean model, to account, for instance, for the spread of heat by ocean currents. In addition, the representation of the energy parks by four model grid boxes (grid spacing is about 3° in latitude and longitude), the smallest area over which the waste heat could be added in the model, is unrealistic in comparison to the likely size of any such installations. In

Table 4.5 The combination of energy parks and heat input in five GCM sensitivity experiments. Source: Williams *et al.* (1979)

EX	Energy parks	Heat input
01	A and C	1.5×10^{14} W at each park
02	B and C	1.5×10^{14} W at each park
03	A only	1.5×10^{14} W
04	A and C	0.75×10^{14} W at each park
05	A and C	1.5×10^{14} W at each park

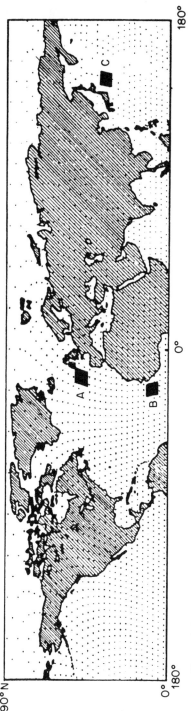

Figure 4.4   Locations of heat input in the GCM simulations of Williams *et al.* (1979), for which results are shown in Figure 4.5

four of the perturbed simulations the heat was added directly to the atmosphere in sensible heat form, by adding 375 W m^{-2} (or 187.5 W m^{-2}).

Figure 4.5 shows the geographical distribution of the difference in 40-day mean sea-level pressure between the energy-park experiments and the average of the three control cases. Shaded areas on the maps show where the ratio of the difference to the standard deviation of the three control cases is greater than 5.0. As Williams *et al.* (1979) explain, this ratio can be considered a 'signal-to-noise' ratio. As discussed in Chapter 2, an important aspect of analysing the results of GCM prescribed change experiments is to determine how much of the difference from control integrations is due to the prescribed change (in the case of this study, the added waste heat) and how much is a result of the model's inherent variability. The latter has been estimated by computing the standard deviations of 40 day means from the three control cases. The ratio of the difference to this standard deviation has a Students' *t* distribution. When the ratio is greater than 5.0 at an individual grid point, there is a 95% chance that the difference is significant and caused by the prescribed change. There are obvious limitations to this statistical approach, particularly since only three control integrations are used, so that much emphasis was placed on examining the similarities in response in experiments that had common features. A consistent response in the vicinity of park A in difference experiments, for example, gives confidence in attributing significance to the results of the experiments.

Returning then to Figure 4.5, we see that in EX1 (Figure 4.5a) there are several large coherent areas of pressure change. There is a sea-level pressure increase over and upstream of park A and a decrease over and downstream of park C on a much larger spatial scale than the area of the energy parks. Pressure changes also occur elsewhere in the hemisphere. The changes in sea-level pressure in EX2, which considered heat input at parks B and C, are neither as large nor as coherent in middle and high latitudes. In none of the energy-park experiments was there a large or significant change in the sea-level pressure fields in tropical latitudes. In EX2, there is no large-scale sea-level pressure change in the vicinity of the energy parks, as there is in EX1.

In EX3, which considered only park A, the magnitude of the sea-level pressure change in the vicinity of the mid-Atlantic energy park is not exactly the same as in EX1, but the sign is the same. That is, in both EX1 and EX3 there is a pressure increase over and upstream of the park and a decrease immediately downstream. Further downstream the two experiments differ, since EX3 has no Pacific energy park. The sea-level pressure changes in EX4 are not similar to those in EX1, which considered the same energy parks but twice as much heat input. Large coherent areas of significant pressure change do not occur in EX4, and in the vicinity of the parks themselves the sea-level pressure changes do not exhibit the characteristics found in EX1, except for the increase in sea-level pressure upstream from the Atlantic park.

The changes in sea-level pressure in EX5 bear some similarities to those in EX1 in the vicinity of the energy parks, with a pressure increase over and upstream of park A and a pressure decrease over and downstream of park C. However, the

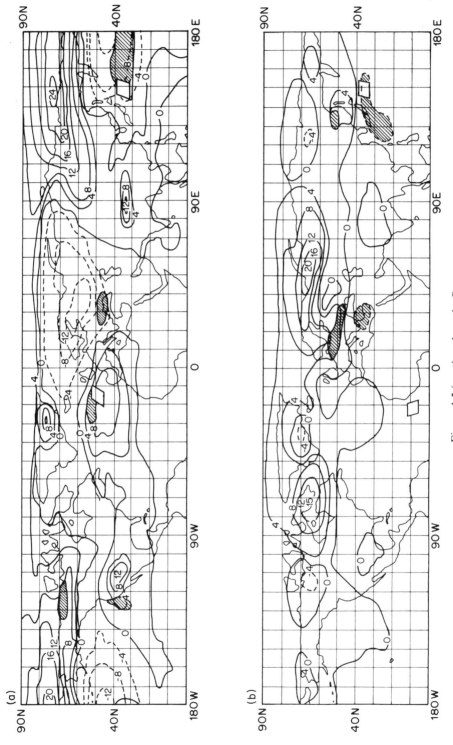

Figure 4.5 (continued overleaf)

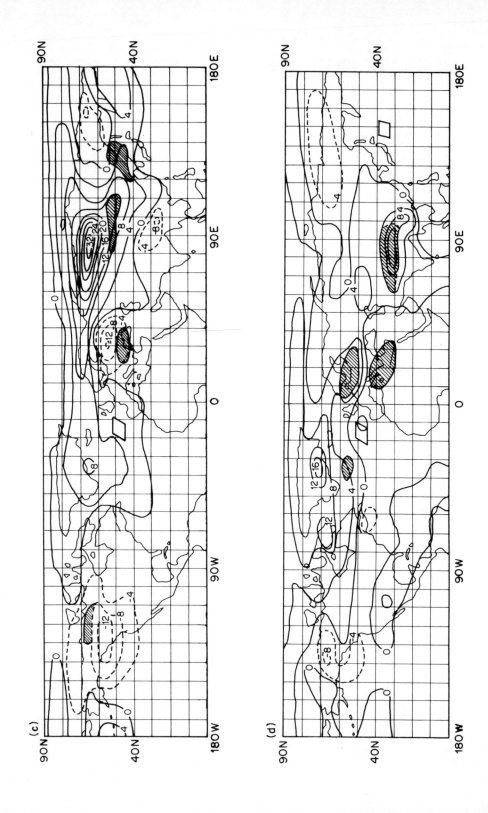

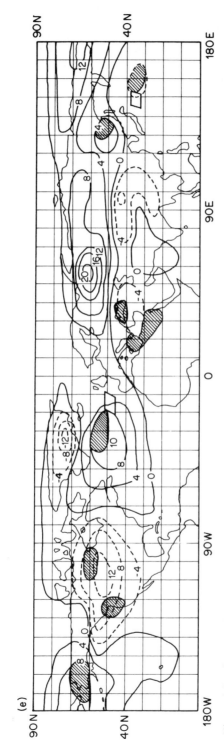

(e)

Figure 4.5   Geographical distribution of the differences in 40-day mean sea-level pressure between energy-park experiments and the average of the three control cases. Shaded areas show where the 'signal-to-noise' ratio is greater than 5.0 based on an estimate of model variability using 40-day means. (a) EX1, (b) EX2, (c) EX3, (d) EX4, (e) EX5. Units: mbar. Source: Williams and Krömer (1979)

large pressure decrease over the Pacific in EX1 (>14 mbar) is not found in EX5 and the pressure decrease downstream from park A in EX1 is also not seen in EX5. The response over North America in the two experiments also differs.

With regard to the distribution of the signal-to-noise ratio, for each of the experiments there are areas where the ratio is large enough to suggest that the results differ significantly from the control-case ensemble. However, as suggested above, the limited number of control cases means that the ratio cannot be used as a definite test of the significance of the waste-heat experiments. In EX1, EX4, and EX5, which each considered parks A and C, the signal-to-noise ratio is greater than 5.0 upstream and downstream of park A, providing statistical evidence supporting the view that the park produces a genuine (and similar in each experiment) response in the model.

This analysis of the sea-level pressure changes in the five waste-heat experiments made with the UK Meteorological Office GCM shows the magnitudes of the responses of one meteorological variable to the input of a large amount of waste heat at energy parks. The responses of other variables, such as the temperature of the lowest atmospheric level, the height of the 500 mbar surface, and the precipitation rate have also been analysed. Overall, the experiments indicate that there is a possibility that the input of unrealistically large amounts of heat on a local scale will cause large, coherent changes in the atmosphere, not just over the areas of heat input but also elsewhere in the hemisphere. The response may vary according to the location, amount, and manner of heat input.

The results of the model experiments must be viewed, however, in terms of both model and scenario shortcomings. The most significant model shortcoming is the absence of a coupled ocean circulation model. In addition, however, the treatment of clouds and hydrological processes is also inadequate. The amounts of heat used so far in model studies have been unrealistically large (150 TW or 300 TW in the experiments with the Meteorological Office GCM compared with a present global energy consumption of less than 10 TW). Input of more realistic amounts of waste heat is likely only to affect local or regional climate rather than global climate.

The latter remarks are supported by Egger (1979), who used two linear-standing wave models to examine the effects of energy parks and extended heat sources. Near the energy parks, Egger finds a response similar to that in the above GCM simulations, i.e., low pressure downstream and high pressure upstream. However, there was almost no agreement in other areas, where the linear model predicts rather weak responses. Egger concludes that energy parks with large heat releases of the order of 10^{14} W would have a noticeable effect on the atmospheric circulation and climate but that a heat input of say 10^{12} W would have no appreciable effect on mesoscale and planetary flow. Lastly, Egger found the response to heat input with a larger horizontal extent had about the same pressure response near the surface as that found with the energy parks.

Williams *et al.* (1979) investigated two methods of waste heat disposal: putting the sensible heat directly into the atmosphere, or into the top 10 m of the vicinity of the energy park. Häfele (1977) suggested that releasing large amounts of waste heat in ocean areas, if done 'intelligently', would have no global climatic impact. For example, he suggests that deep cold ocean water could be pumped to the surface and used to dispose of waste heat with no resultant change in ocean surface temperatures. Baker (1977) has discussed this possibility in some more detail. A park with an energy input of 40 GW, covering an area of 25 km² and dissipating the heat in a 100 m layer of ocean water, would lead to a local temperature increase of 1 °C in about 3–10 days, depending on the rate of flow of the water. If, on the other hand, local heating is avoided by using deep cold water as coolant, then the surface temperature could be maintained but, as Baker points out, the mass and hence the total heat transport would be changed. Obviously more quantitative modelling is required on several aspects of energy parks in ocean areas.

4.7   THE EFFECT ON GLOBAL CLIMATE OF HEAT RELEASE FROM MEGALOPOLITAN AREAS

As with the studies reported in the last section, the effect of heat releases from megalopolitan areas has been investigated by using models of the atmospheric circulation. Simulations with the NCAR GCM, to investigate the effects of heat release from a megalopolitan area in the eastern United States, have been reported by Llewellyn and Washington (1977) and Washington and Chervin (1979). Figure 4.6 shows the area over which the heat was added. The amount of heat added was 90 W m⁻², i.e., that presently used in Manhattan. Other regions of the globe were not modified.

Llewellyn and Washington (1977) discuss the results of a January simulation with this heat release. Figure 4.7 shows the difference in temperature at

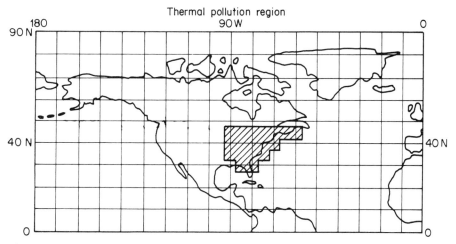

Figure 4.6   Location of heat input (90 W m⁻²) in GCM simulations by Llewellyn and Washington (1977) and Washington and Chervin (1979). Source: Washington and Chervin (1979)

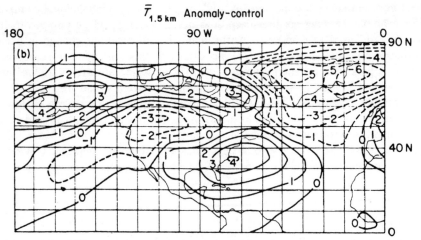

Figure 4.7 Difference in temperature at 1.5 km between control simulation (January, without heat input) and simulation with heat input as in Figure 4.5. Temperatures are averages from days 31–60 of each simulation. Source: Llewellyn and Washington (1977)

anemometer level between the control and heat-release cases. The heat release results in a temperature increase of up to 12 °C in the vicinity of the heat input, but no other location shows such a large response. Llewellyn and Washington concluded that the changes due to the heat release are large in the boundary layer in the vicinity of the heat input but are not significant above this layer.

Washington and Chervin (1979) describe further integrations with a different version of the NCAR GCM to investigate the response to the megalopolis scenario described above, both January and July experiments being performed. In the region where the heat was added, the waste heat produced large and statistically significant changes in the mean temperature, vertical velocity, precipitation, and soil-moisture fields. However, considerable differences in the model response in different seasons were noted. The temperature increase in the lower troposphere in the January simulation was much larger than in the July simulation. They suggest that this is because the atmosphere is more stable in January and because the added heat is a much larger component of the surface energy budget in the winter case. The maximum temperature change at the ground in the megalopolis region was 12 °C in January and 3 °C in July. However, the temperature changes in the lower troposphere were smaller than those at the surface in January by a factor of 4, but only smaller by a factor of 2 in July, suggesting that the heat is more vertically mixed in the summer case. The precipitation tended to increase over the megalopolis, but, since evaporation also increased because of the larger surface temperature, the soil-moisture amounts decreased. The effects of the megalopolis heat release were, however, localized, although in the July experiment the heat island extended to the west of the prescribed heat input. No statistically significant globally averaged changes were found for any temperature or precipitation field in either the January or July case.

Krömer *et al.* (1979) have described a set of simulations made with the UK Meteorological Office GCM to investigate the effect of heat release from megalopolitan areas. In this study six regions of heat release were considered, each of them representing areas where a large population and/or energy consumption density could be expected in the future. Figure 4.8 shows the locations of the six regions where heat was released. In the first simulation the total heat release from the six regions was 300 TW, while in the second and third simulations the heat releases were 50 TW and 30 TW respectively. The size of heat release areas and their heat input were selected so that the heat released per square metre was the same for each grid point. The simulations were all made using January boundary conditions.

Figure 4.9 shows the geographical distribution of the differences in 40 day mean sea-level pressure between the megalopolis simulations and the average of three control cases. The shading indicates areas where the difference is five times greater than the standard deviation of the sea-level pressure in the three control cases—a measure of significance. In the 300 TW case it is seen that large coherent changes of sea-level pressure occur, and not only over the areas of heat release. For example, a large change occurs over an extended area covering eastern Siberia and most of Canada, its maximum being a 16 mbar pressure increase over Alaska. It is also notable that, in almost all of the six areas of heat input, large regional pressure decreases occurred in this experiment. The addition of 300 TW is the same input as that described above for ocean-energy-park simulations. The magnitude of the changes in the energy-park simulations and the megalopolis simulation are comparable, although the magnitude and locations of the individual changes differ. The changes of mean sea-level pressure in the 50 TW megalopolis case are similar to those in the 300 TW case, although the extent of the shaded areas is somewhat smaller. Thus, although the heat input has been decreased from 300 to 50 TW, the response is not proportionately reduced; this illustrates well the non-linearity of the system. When the heat input is reduced to 30 TW, the response is smaller and probably not significant.

An examination of the differences in 40 day mean temperature of the lowest atmospheric layer showed that in the first simulation large changes occurred mainly over the continents. Away from the heat input areas, the temperature response appeared to be forced by the pressure response. For example, there was a cooling over Canada and Siberia related to the increased pressure there. Over the areas of heat input a large regional temperature increase generally occurred. Similar observations were made for the 50 TW case, where the changes occurred mainly over the continents and were positive over most of the areas of heat input. In both simulations there was a temperature decrease downstream of the European megalopolis, which was larger than any increase due to heat input from the USSR megalopolis, so that a temperature decrease of 4–8 °C over the latter occurred in both experiments. This illustrates the potential interactions of more than one area of heat input. In the 30 TW case the temperature changes were smaller and again probably not significant.

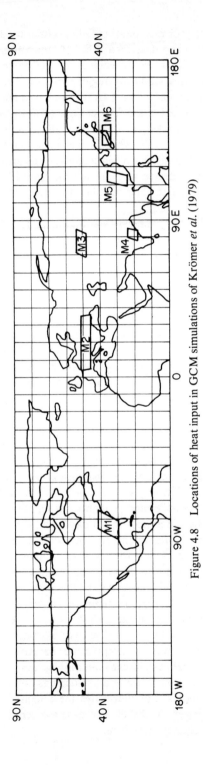

Figure 4.8   Locations of heat input in GCM simulations of Krömer *et al.* (1979)

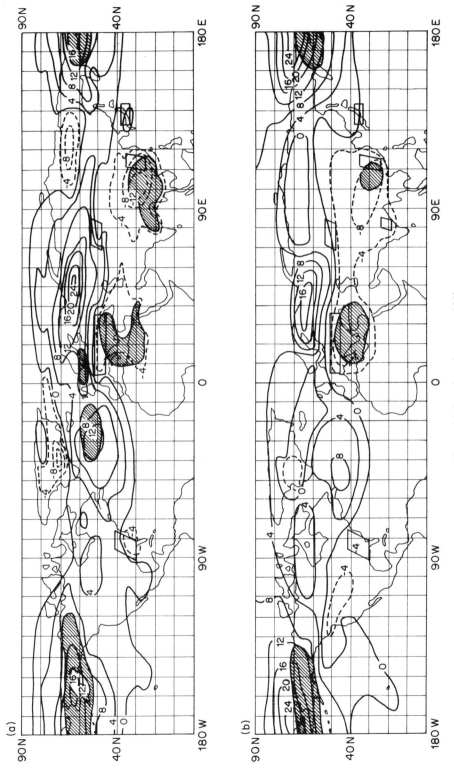

Figure 4.9 (continued on page 138)

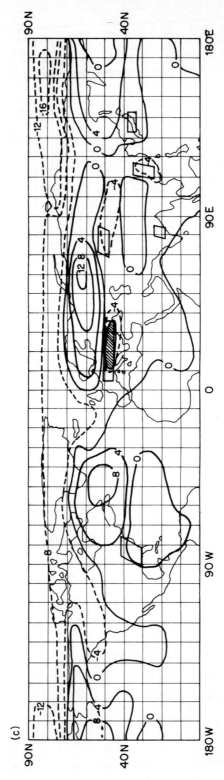

Figure 4.9 Differences in 40-day means of sea-level pressure between the average of three control cases and simulations with heat input at locations indicated in Figure 4.7. Total heat input was (a) 300 TW, (b) 50 TW, (c) 30 TW. Shaded areas show where the differences are five times greater than the standard deviation of the three control cases. Units: mbar. Source: Krömer et al. (1979)

Therefore, it has been found that, when a heat input of 300 TW is spread over six continental areas, the hemispheric response is comparable in magnitude to that when the heat input is concentrated at only two ocean energy parks. There appears to be a significant model response to the heat input, including a strong regional response in some of the megalopolis areas and large coherent changes elsewhere. If the heat input is reduced to 50 TW, the effect decreases only to a small extent, emphasizing the non-linear behaviour of the atmosphere. A further reduction of the waste heat release to 30 TW, however, seems to generate a model response that is not larger than the inherent model variability, as determined from the model control cases.

Thus model studies of the potential influence of megalopolitan areas on the atmosphere suggest that an extremely large amount of heat input could influence the atmospheric circulation to such an extent that the regions of heat input will not be the only ones experiencing changes in variables such as sea-level pressure, temperature, and rainfall. With heat inputs of the order of magnitude expected during the next century, no significant hemispheric response in the model was determined, but large regional changes over the areas of heat input were observed.

REFERENCES

Atkinson, B. W. (1968). A preliminary examination of the possible effect of London's urban area on the distribution of thunder rainfall, 1951–60. *Trans. Inst. Brit. Geogr.*, **44**, 97–118.

Bach, W. (1979). Waste heat and climatic change. In, L. Theodore, A. J. Buonicore, and E. J. Rolinski (eds.), Volume I: *Perspectives on Energy and the Environment.* CRC Press, West Palm Beach Florida, USA.

Baker, D. J. (1977). Ocean dynamics and energy transfer: Some examples of climatic effects. In, *Energy and Climate.* National Academy of Sciences, Washington, DC.

Barry, R. G., and R. J. Chorley (1968). *Atmosphere, Weather and Climate.* Methuen, London.

Bhumralkar, C. M., and J. Alich (1976). Meteorological effects of waste heat rejection from power parks. *Power Engr.*, 57–61.

Bhumralkar, C. M., and J. Williams (1981). *Atmospheric Effects of Waste Heat Discharge.* Marcel Dekker, New York.

Bhumralkar, C. M., J. Williams, and A. Slemmons (1979). The impact of a conceptual thermal electric conversion plant on regional meteorological conditions: A numerical study. *Solar Energy*, **23**, 393–403.

Briggs, G. A. (1974). Plume rise from multiple sources. *Proceedings of 'Cooling Tower Environment 1974' Symposium.* University of Maryland, USA.

Briggs, G. A. (1975). Plume rise predictions. Lectures on Air Pollution and Environmental Impact Analyses. American Meteorological Society, Boston, Massachusetts, USA.

Budyko, M. I. (1969). The effect of solar radiation variations on the climate of the earth. *Tellus*, **21**, 611–619.

Burwell, C. C., M. J. Ohaman, and A. M. Weinberg (1979). A siting policy for an acceptable nuclear future. *Science*, **204**, 1043–1051.

Chandler, T. J. (1965). *The Climate of London*, Hutchinson, London.

Changnon, S. A. (1973). Recent studies of urban effects on precipitation. In, *Proc. Workshop on Inadvertent Weather Modification.* Utah State University, Logan, Utah, Aug. 1973, 111–139.

Changnon, S. A. (1980). More on the La Porte anomaly: A review. *Bull. Amer. Meteor. Soc.*, **61**, 702–711.

140

Church, C. R., J. T. Snow, and J. Dessens (1980). Intense atmospheric vortices associated with a 1000 MW fire. *Bull. Amer. Meteor. Soc.*, **61**, 682–694.

Culkowski, W. M. (1962). An anomalous snow at Oak Ridge, Tennessee. *Mon. Wea. Rev.*, **90**, 194–196.

Egger, J. (1979). The impact of waste heat on the atmospheric circulation. In, W. Bach, J. Pankrath, and W. W. Kellogg (eds.), *Man's Impact on Climate*. Elsevier, Amsterdam.

Everett, R. G., and Zerbe, G. A. (1977). Winter field program at the Dresden cooling ponds. ANL 76-88, Part IV, Radiol. and Envir. Res. Div. Ann. Rept., *Atmospheric Physics*, Jan.–Dec. 1976, 108–113.

Fortak, H. G. (1979). Entropy and climate. In, W. Bach, J. Pankrath, and W. W. Kellogg (eds.), *Man's Impact on Climate*. Elsevier, Amsterdam.

Garstang, M., P. D. Tyson, and G. D. Emmitt (1975). The structure of heat islands. *Rev. Geophys. Space Phys.*, **13**, 139–165.

Geiger, R. (1965). *Climate Near the Ground*, Harvard University Press, Cambridge, Massachusetts.

Häfele, W. (1977). Das Klima als Randbedingung für globale Energiesysteme. In, K. Strand and H. Porias (eds.), *Grosstechnische Energienutzung und Menschlicher Lebensraum*. International Institute for Applied Systems Analysis, Laxenburg, Austria.

Hanna, S. R. (1974). Fog and drift deposition from evaporative cooling towers. *Nuclear Safety*, **15**, 190–196.

Hanna, S. R. (1978). Effects of power production on climate. American Meteorological Society Conference on Climate and Energy: Climatological Aspects and Industrial Operations, 8–12 May 1978, Asheville, North Carolina.

Hanna, S., and E. Gifford (1975). Meteorological effects of energy dissipation of large power parks. *Bull. Amer. Meteor. Soc.*, **56**, 1969–2076.

Hosler, C. L., and Landsberg, H. E. (1977). The effect of localized man-made heat and moisture sources in mesoscale weather modification. In, *Energy and Climate*. National Academy of Sciences, Washington, DC.

Kellogg, W. W. (1977). Effects of human activities on global climate. WMO Techn. Note No. 156, WMO-No. 486, World Meteorological Organization, Geneva.

Koenig, L. R., and C. M. Bhumralkar (1974). On possible undesirable atmospheric effects of heat rejection from large electric power centers. R-1628-RC, Rand Corporation, California.

Kramer, M. L., D. E. Seymour, M. E. Smith, R. W. Reeves, and T. T. Frankenberg (1976). Snowfall observations from natural draft cooling tower plumes. *Science*, **193**, 1239–1241.

Krömer, G., J. Williams, and A. Gilchrist (1979). Impact of waste heat on simulated climate: A megalopolis scenario. WP-79-73, International Institute for Applied Systems Analysis, Laxenburg, Austria.

Landsberg, H. E. (1975). Man-made climatic changes. In S. F. Singer (ed.), *The Changing Global Environment*. Reidel, Dordrecht, Holland, pp. 197–234.

Landsberg, H. E. (1977). Rainfall variations around a thermal power station. *Atmos. Environ.*, **11**, 565.

Landsberg, H. E. (1979). The effects of man's activities on climate. In, M. R. Biswas and A. K. Biswas (eds.), *Food, Climate and Man*, J. Wiley and Sons, New York.

Landsberg, H. E. (1981). *The Urban Climate*. Academic Press.

Leahey, D. M., and J. P. Friend (1971). A method of predicting the depth of the mixing layer over an urban heat island with application to New York City. *J. Appl. Meteor.*, **10**, 1162–1173.

Llewellyn, R. A., and W. M. Washington (1977). Regional and global aspects. In, *Energy and Climate*. National Academy of Sciences, Washington, DC.

Martin, A. (1974). The influence of a power station on climate—a study of local weather records. *Atmos. Environ.*, **8**, 419–424.

Matson, M., E. P. McClain, D. F. McGinnis, Jr., and J. A. Pritchard (1978). Satellite detection of urban heat islands. *Mon. Wea. Rev.*, **106**, 1725–1734.

Munn, R. E., and L. Machta (1979). Human activities that affect climate. *Proc. of the World Climate Conference.* WMO Publication No. 357, World Meteorological Organization, Geneva.

Murphy, A. H., A. Gilchrist, W. Häfele, G. Krömer, and J. Williams (1976). The impact of waste heat release on simulated global climate. RM-76-79, International Institute for Applied Systems Analysis, Laxenburg, Austria.

Oke, T. R. (1974). Review of urban climatology, 1968–1973. WMO Tech. Note No. 134, World Meteorological Organization, Geneva.

Oke, T. R. (1980). Climatic impacts of urbanization. In, W. Bach, J. Pankrath, and J. Williams (eds.), *Interactions of Energy and Climate.* Reidel, Dordrecht, Holland.

Oke, T. R. (1982). The energetic basis of the urban heat island. *Quart. J. Roy. Meteor. Soc.*, **108**, 1–24.

Oke, T. R., and C. East (1971). The urban boundary layer in Montreal. *Boundary Layer Meteorol.*, **1**, 411–437.

Oke, T. R., and F. G. Hannell (1970). The form of the urban heat island in Hamilton, Canada. In, Urban climates. WMO Techn. Note 108, World Meteorological Organization, Geneva.

Ott, R. E. (1976). Locally heavy snow downwind from cooling towers. NOAA Tech. Memo, NWSER-62, WSFO, Charleston, West Virginia.

Penner, S. S. (1976). Monitoring the global climatic impact of direct heat addition associated with escalating energy use. *Energy*, **1**, 407–412.

Rotty, R. M. (1974). Waste heat disposal from nuclear power plants. NOAA Tech. Memo, ERL ARL-47, NOAA, Environmental Research Laboratories, 28 pp.

Rotty, R. M., H. Baumann, and L. L. Bennett (1976). Atmospheric considerations regarding the impact of heat dissipation from a nuclear energy center. ORNL/TM 5122, Oak Ridge National Laboratory, 39 pp.

Sawyer, J. S. (1965). Notes on the possible physical causes of long-term weather anomalies. WMO Techn. Note No. 65, World Meteorological Organization, Geneva.

Schmaus, A. (1927). Großstadt und Niederschlag. *Meteor. Z.*, **64**, 339–341.

Schneider, S. H., and R. D. Dennett (1975). Climate barriers to long-term energy growth. *Ambio*, **4**, 65–74.

Sellers, W. D. (1969). A global climate model based on the energy balance of the earth–atmosphere system. *J. Appl. Meteor.*, **8**, 329–340.

SMIC (1971). *Inadvertent Climate Modification.* Report of the study of man's impact on climate, MIT Press, Cambridge, Massachusetts.

Taylor, F. G., L. K. Mann, R. C. Dahlman, and F. L. Miller (1974). Environmental effects of chromium and zinc in cooling-water drift. ERDA Symposium Series. CONF-740302, NTIS, Springfield, Virginia.

Taylor, R. J., S. T. Evans, N. K. King, E. T. Stephens, D. R. Packham, and R. G. Vines (1973). Convective activity above a large-scale brushfire. *J. Appl. Meteor.*, **12**, 1144–1150.

Washington, W. M. (1971). On the possible uses of global atmospheric models for the study of air and thermal pollution. In, W. H. Matthews *et al.* (eds.), *Man's Impact on Climate*, MIT Press, Cambridge, Massachusetts, pp. 265–276.

Washington, W. M. (1972). Numerical climatic change experiments: The effect of man's production of thermal energy. *J. Appl. Meteor.*, **11**, 763–772.

Washington, W. M., and R. M. Chervin (1979). Regional climatic effects of large-scale thermal pollution: Simulation studies with the NCAR general circulation model. *J. Appl. Meteor.*, **18**, 3–16.

Weber, M. R. (1978). Seasonal variations in temperature in the vicinity of two nuclear power plants: A comparison of operational and preoperational data. American Meteorological Society Conference on Climate and Energy: Climatological Aspects and Industrial Operations. 8–12 May 1978, Asheville, North Carolina.

Wetherald, R. T., and S. Manabe (1975). The effects of changing the solar constant on the climate of a general circulation model. *J. Atmos. Sci.*, **32**, 2044–2059.

Williams, J., and G. Krömer (1979). A systems study of energy and climate. SR-79-2B. International Institute for Applied Systems Analysis, Laxenburg, Austria.

Williams, J., G. Krömer, and A. Gilchrist (1977a). Further studies of the impact of waste heat release on simulated global climate: Part 1. RM-77-15, International Institute for Applied Systems Analysis, Laxenburg, Austria, 22 pp.

Williams, J., G. Krömer, and A. Gilchrist (1977b). Further studies of the impact of waste heat release on simulated global climate: Part 2. RM-77-34, International Institute for Applied Systems Analysis, Laxenburg, Austria, 33 pp.

Williams, J., G. Krömer, and A. Gilchrist (1979). The impact of waste heat release on climate: Experiments with a general circulation model. *J. Appl. Meteor.*, **18**, 1501–1511.

CHAPTER 5

The Effect on Climate of Particles and Gases of Man-made Origin (Excluding CO_2)

5.1 INTRODUCTION

In addition to releasing waste heat and/or CO_2 into the atmosphere, energy systems can also release other gases and particles that can influence the climate. Combustion of fossil fuels releases gases, soot, and ash particles directly into the atmosphere and particles (and aerosols) are also formed by chemical reactions within the atmosphere from the gaseous products of combustion, e.g., sulphates, organic nitrates, sulphuric and nitric acids, and hydrocarbons. Other gases and particles are released into the atmosphere as a result of human activities, for instance, chlorofluorocarbons and oxides of nitrogen, and can influence climate either by their greenhouse effect or by affecting the concentrations of other climatically important trace gases in the atmosphere, especially ozone. These topics will be addressed in more detail in this chapter.

5.2 THE EFFECT OF PARTICLES ON CLIMATE

(a) Sources of particles

It has been established from theory and observation (e.g., Mitchell, 1975) that particles with a diameter of 0.1–5 μm are significant for the heat balance of the earth–atmosphere system, because they are relatively abundant in the atmosphere and are effective in scattering, absorbing, and thereby attenuating, solar radiation. Particles smaller than 0.1 μm have negligible mass and larger particles do not have a long enough residence time in the atmosphere to have more than a localized effect on climate.

Table 6.3 in the next chapter gives one set of estimates of particle production at the present day, based on data from Dittberner (1978). Table 5.1 compares a number of estimates of the total global particle release to the atmosphere. It is clear that there is considerable uncertainty regarding the magnitude of both natural and anthropogenic particle release. In all cases, the global particle release due to man's activities is smaller than the total release, but it is not an insignificant fraction. It is estimated that about 90% of the particles in the atmosphere at the

Table 5.1 Global particle release to the atmosphere

Author	Natural source (10^{12} g/yr)	Anthropogenic (10^{12} g/yr)
Dittberner (1978) (<5 μm)	1,150–1,498	692
Robinson (1977) (all particle sizes)	2,312 (range 773–2,200)	296 (range 185–415)
Mitchell (1975) (<5 μm) (all particle sizes)	1,100 2,065	490 785
Bach (1979) (all particle sizes)	3,000	300
Landsberg (1979) (all particle sizes)	2,000	~400
Ellsaesser (1975) (all particle sizes)	3,049	396

present are confined to the troposphere. The other 10% are in the stratosphere and consist primarily of volcanic dust. The stratospheric loading may vary by as much as two orders of magnitude from year to year because it depends on the timing and strength of significant volcanic eruptions.

Changes in the particle emission rates due to human activities as a function of time have been discussed by Munn and Machta (1979), who point out that the use of more efficient methods of combustion, cleaner fuels, smoke removal equipment, and tall chimneys has led to a steady improvement in air quality in many towns in recent decades. However, although the black smoke above and downstream of urban-industrial areas has disappeared in Europe and North America, it has been replaced by brown haze which sometimes extends over very large areas. Munn and Machta (1979) state that urban emissions of photochemically active gases and hydrocarbons have been increasing, leading to increases in sulphate, nitrate, and phosphate particles, sometimes 500 to 1,000 km from the source regions.

The question of whether the particle concentration of the atmosphere is increasing is open and very controversial. Ellsaesser (1975) cited a considerable amount of evidence showing that local sources of particles in industrialized regions are being brought under control and that there was no global upward trend due to anthropogenic sources. Figure 5.1 shows a time series of the particulate load at Kew (London) between 1922 and 1971. It shows a strong downward trend. Landsberg (1979) notes that other localities have shown measurable but not as spectacular results, while still other localities have a static dust load. Increased industrialization and automobile traffic have, according to Landsberg, balanced decreases resulting from control efforts, and the agricultural contribution has remained essentially uncontrolled.

Naegele and Sellers (1981) have reported the results of a study of airport visibility data for the period 1958–1978 (or 1979, depending on data availability) for 18 cities in the western and southwestern United States. They found no general trends over the total period. However, visibilities generally decreased from 1958 to 1972 and generally increased from 1973 to 1978. The authors suggest that this

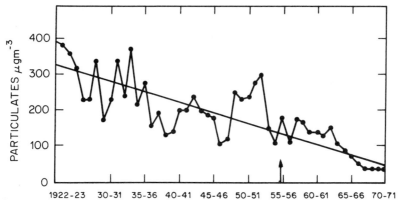

Figure 5.1 Time series of atmospheric particulate load at Kew near London, England. Arrow marks introduction of controls (Clean Air Act) (after Aulicierns and Burton, 1973)

reflects the positive effects of the implementation of the Clean Air Act of 1973. At Yuma, Tucson, and Phoenix there was a high correlation between visibility and emission of sulphur oxides from local copper smelters.

Landsberg (1979) points out that the aerosols produced by high-flying aircraft could have significant effects. Carbon particles result from incomplete combustion of fuel. Sulphur dioxide in fuel is transformed to sulphuric acid that is in solid form at low stratospheric temperatures. As Landsberg points out, the main reason for concern is because these particles and aerosols are only transported out of the stratosphere slowly and thus have a long residence time, particularly in the higher stratospheric layers. Large increases in aerosols and particles in the upper atmosphere would probably lead to a global cooling. Aircraft and spacecraft have so far contributed only minor quantities of aerosols and particles to the stratosphere and are therefore not climatically significant up to the present (Pollack *et al.*, 1976).

(b) The effect of particles on the radiation balance

The effects of particles on the atmosphere–earth radiation balance depend on a number of factors, which have been listed by Munn and Machta (1979) as:

(i) the size distributions of the particles;
(ii) the shapes of the particles;
(iii) the scattering, reflection, and radiative properties of the particles;
(iv) the vertical distributions of particles;
(v) the relative humidity;
(vi) the time variabilities of these factors.

An increase in the particle loading of the atmosphere can lead to:

(i) a change in the amount of solar radiation scattered back to space (causing a change in the mean planetary temperature);
(ii) more absorption of solar radiation in the atmosphere (causing an increase of atmospheric temperatures);
(iii) less solar radiation reaching the earth's surface (causing a decrease of earth surface temperature);
(iv) an increase in the ratio of diffuse to direct solar radiation, which has been shown experimentally to lead to an increase in surface albedo and thus to surface cooling.

The net effect of particles on the radiation balance of the earth–atmosphere system is very difficult to assess. Kellogg (1980) has reviewed the physics involved and shown how the theory can be applied using available data on the optical properties of particles. The mathematical treatment of scattering and absorption by particles has been given by Coakley and Chylek (1975), Viskanta et al. (1977), and Atwater (1975). Kellogg (1980) summarizes the results of analyses of the effect of particles on the albedo of the surface plus lower atmosphere (the effective albedo). Over land, particles cause the effective albedo to be less than the surface albedo and therefore the effect of the particles is to warm the lower atmosphere. According to Kellogg, over the oceans, in clear air the particles may cause cooling, but in cloudy areas they also cause warming.

Eiden (1979) has also shown that the amount of radiation scattered and absorbed depends on the albedo of the underlying surface and the absorption characteristics of the particles. Figure 5.2, from Eiden (1979), illustrates this point. We see that, in the case of both absorbing and non-absorbing particles, the solar energy reaching the earth's surface will be reduced. The atmosphere obviously absorbs more radiation if absorbing particles are added and less radiation if non-absorbing particles are added. The earth–atmosphere system absorbs less radiation when non-absorbing particles are added; when absorbing particles are added, it absorbs less when the planetary albedo is low and more when the planetary albedo is high (greater than about 35%). Therefore, the addition of absorbing particles will cause a redistribution of the solar energy absorbed by the planet, with more absorbed by the atmosphere and less by the earth's surface. All of this assumes that the original cloud cover and water-vapour content of the atmosphere do not change. Moreoever, only an initial thermodynamic change has been considered; the change in vertical temperature gradient would lead to further atmospheric changes. Thus, the effect can only be assessed realistically using an atmospheric circulation model. Furthermore, Eiden (1979) has pointed out that increases in the particle content of the atmosphere would produce latitudinally varying effects. An increase in the particle content in middle latitudes would lead to an intensified horizontal temperature gradient between the mid-latitudes and the pole in the lower atmosphere and to a reduced temperature gradient at the surface. A similar increase in particle loading in the polar and/or tropical regions

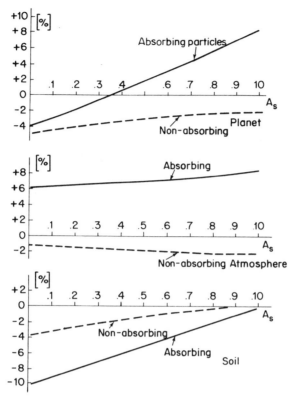

Figure 5.2 Relative change, due to a doubling of the concentration of particles, of solar energy absorbed by the earth–atmosphere system (above), the atmosphere (centre), and the earth's surface (below). Source: Eiden (1979)

would cause only second- or higher-order effects on the solar radiation transfer: in the polar atmosphere particles are not of significance in the radiation balance; in the tropical atmosphere the effect of water vapour dominates that of particles. The final effect of such horizontal temperature gradient changes can also only be evaluated with the aid of a dynamic model.

Grassl (1979) has discussed the influence of particles on the planetary albedo in cloud-free and cloudy areas; depending on surface albedo and the imaginary part of the refractive index of the particles, an increase in particles can lead to an increase or decrease of albedo with the crossing point virtually independent of particle size and optical depth. Within cloudy areas, Grassl distinguishes between three effects that change the albedo. The combination of these effects means that an increase in the particles enhances cloud albedo for thin clouds and decreases cloud albedo for thick clouds. However, it is pointed out that an estimate of the total planetary albedo change due to particle increases is not possible at present because of lack of data on the imaginary part of the refractive index of particles and on the mean optical depth of clouds.

Although many theoretical studies have shown that aerosol-induced changes in the earth–atmosphere albedo could have significant effects on climate, there has been a lack of documentation of albedo changes caused by actual aerosol layers. Russell *et al.* (1979) have, however, reported a case in which the measured earth–atmosphere albedo was increased by about 0.01 (from 0.11 to 0.12) by a transient aerosol layer. The measurements were made in San Francisco with instruments at three levels on a television tower. The albedo change associated with an increase in aerosol concentration occurred for about an hour. Russell *et al.* (1979) point out that the aerosol event was very short-lived and its areal extent is not known, so that the climatic significance of this event is virtually nil. The analysis is, however, very useful, since it provides a method for incorporating the optical properties of actual aerosol layers into a flexible model and computing albedo changes. Moreover, the authors point out that aerosol layers with an optical thickness equal to or greater than that at the peak of the observed event are rather common in populated regions of the globe.

It is clear that the effects of increased particle loading on the radiation balance of the earth–atmosphere system are extremely complex, depending on the cloud characteristics and on the characteristics of the particles and the location of the particles with respect to the underlying surface. Moreover, the changes in heating rates would produce changes in horizontal and vertical temperature gradients, with resulting changes in cloudiness and resulting radiative effects. All of these can ultimately only be assessed by using a complex numerical model of the entire earth–atmosphere system, for which detailed information on particle characteristics, presently not available, would be required. Early studies suggested that an increased particle loading would increase the scattering of solar radiation back to space, giving a net cooling of the earth–atmosphere system. Recent studies have suggested that the particles lead to a warming. For example, Kellogg (1977) suggested that evidence shows that most of the anthropogenic particles exist over land, where they reduce the net earth–atmosphere albedo and therefore warm the system. However, as was discussed in Chapter 3 with regard to CO_2 increases, the regional changes of atmospheric circulation will be more significant than global average changes in planetary temperature.

Some model studies of the effect on climate have been carried out with simpler models of the climate system. Rasool and Schneider (1971) used a one-dimensional model to show that for a six- to eight-fold increase in particle loading, the mean temperature in the model decreased by 3.5 K. However, Mitchell (1975) points out that these results should be viewed as an upper limit of an expected cooling effect because of assumptions made.

Reck (1974, 1975) has investigated the response of the Manabe and Wetherald (1967) one-dimensional radiative–convective model to adding particles to the atmosphere. Results of one set of experiments suggested that a doubling of the particle loading in the lower layer of the troposphere in polar areas would lead to a surface temperature increase. One-dimensional model experiments by Wang and

Domoto (1974) indicated that heating could occur with increased particle concentration if the surface albedo were sufficiently high (greater than 30%). Weare *et al.* (1974) have shown that the effect of added particles depends on the location of the particles within the atmosphere with respect to average cloud amount, the cloud reflectivity, and the underlying surface reflectivity.

Bryson and Dittberner (1976) developed a simple mean hemispheric temperature model in the form of a differential equation that is a function of the CO_2 content of the air, and volcanic and anthropogenic particle emissions. It was assumed (unrealistically) that the primary effect of particles is to reduce the amount of solar energy that reaches and is absorbed by the earth's surface; thus it was assumed that increased particle loading would lead to a surface cooling. It should be noted here that there is in fact a difference between the effect of anthropogenic particles, which mainly stay in the lower atmosphere, and volcanic particles, which are injected into the stratosphere (Mitchell, 1975). A net cooling of both the earth's surface and the lower atmosphere is to be expected and has been observed following an increase in stratospheric particle loading.

Charlock and Sellers (1980b) have pointed out that radiative–convective models can be used to study the radiative interactions of particles, clouds, gases, and the surface, but that, when they are applied in the usual globally and annually averaged form, the models do not describe the effect of particles on the surface temperature of a particular geographical region or season, which may be more or less sensitive to radiative effects than the globe as a whole. Moreover, the same authors point out that heat storage and meridional transport effects have generally not been included in earlier steady-state radiative–convective studies of particles and climate. Therefore, Charlock and Sellers (1980b) have used a one-dimensional, radiative–convective model with the addition of sources/sinks of heat due to surface and atmospheric storage and to oceanic and atmospheric dynamics. This allows for the calculation of effects of particles for selected latitude belts on a month-by-month basis. The model results of Charlock and Sellers show that the addition to the atmosphere of the presently observed particle loading (optical depth of 0.125; Toon and Pollack, 1976) reduces the global average surface temperature of the order of 1.5 K in comparison with an atmosphere without particles. With ice albedo feedback, one version of the model showed that the present particle loading causes a temperature drop of 3.2 K.

(c) The effect of particles on cloudiness and precipitation

In addition to interacting with the radiation field, particles, especially those produced by burning fossil fuels, incinerating garbage, and running automobiles, act as condensation or freezing nuclei. That is, the particles can initiate the formation of cloud droplets or hasten the freezing of cloud droplets if the temperature is below 0 °C. The particles most commonly produced by combustion of oil or coal, i.e., sulphates, are very efficient condensation nuclei.

As pointed out in Chapter 4, there is evidence that the amount of rainfall over and downwind of cities is greater than in rural areas (see Table 4.4). Not all of this increase can be attributed to an increase in cloud condensation nuclei, since increased convection in the urban heat island could also enhance precipitation. Landsberg (1975) points out that an observed variation of urban precipitation in accordance with the human work week also argues for at least a residual effect of nucleating particles produced in cities. A review by Schaefer (1975) of measurements of particle concentrations near urban areas and their observed effects indicates that particle pollution is modifying cloud patterns over large areas of the globe and influencing precipitation patterns and types. As Kellogg (1977) has emphasized, it is difficult to assess these effects quantitatively even on a regional scale, but they should be recognized.

Further studies of the interactions of particles and clouds have been reported, for example, by Twomey (1972) and Liou (1976), who find that theoretical calculations suggest that clouds should be more reflective than they are actually observed to be. This difference is thought to be due to the presence of particles, and the decrease in cloud reflectivity is thought to occur whether the particles are included within the cloud droplets or floating between them. The observed reduction of cloud reflectivity is 10–20%, so that any increase in absorbing particles could cause additional absorption of radiation by clouds, giving a further source of atmospheric heating and surface cooling due to increased particle loading.

Charlock and Sellers (1980a) have used a one-dimensional radiative–convective model to investigate the effect of changes in aerosol concentration on cloud condensation nuclei (CCN) concentrations and thus on surface temperature. Using Twomey's (1977) calculations of cloud reflectivity as a function of CCN concentration, they investigated the effect of varying low-cloud reflectivity. The results show that a doubling of CCN would increase global low cloud albedo by about 0.045 and reduce surface temperature by 0.9 K. On the other hand, a doubling of the cloudless atmospheric aerosol reduces the model surface temperature by 1.3 K. Thus, they concluded that, if CCN increase in proportion to the increase in other aerosol constituents, the effects due to changes in cloud reflectivity are of the same order of magnitude as the effects on the cloudless parts of the atmosphere. Charlock and Sellers (1980a) point out that aerosol modifications to only one cloud type were considered, and low cloud was selected because the increased aerosol concentrations in the near future will mostly be in the lower troposphere. CCN effects on cloud infrared optics were neglected.

(d) Comments on the climatic effects of particles

The preceding literature review has indicated the complexity of the climatic effects of particles. If only the radiative effects are considered, adding particles can cause warming or cooling, depending on many variables, including the nature of the particles themselves and the location of input. However, since the radiative warming or cooling would change both vertical and horizontal temperature gradients, dynamical effects would also occur. Particles can also cause changes in condensation and precipitation and in the radiative properties of clouds.

It is clear that the consequences of increased atmospheric particle loading can ultimately only be realistically assessed using comprehensive models of the atmosphere–ocean–ice system. Most of the model experiments made to date have used one-dimensional, radiative–convective models and have studied the effect on globally or zonally averaged surface temperature of increased particle loading. However, in addition to the need to incorporate dynamics into such studies, there is a need to study the effects of regional particle loading on regional climate. This is so in particular because regional air pollution is already observed on a significant scale in industrialized regions such as the United States of America and Western Europe. Also, since particles have a comparatively short life in the atmosphere (about a week before they are washed out), the particle concentration increase is likely to have effects on the regional scale. This implies that models are needed that consider dynamics, radiative heating and cooling and the hydrological cycle, including cloudiness, applicable for various boundary conditions on a regional scale. In addition, observations of the detailed characteristics of aerosols and their effects on local energy balance, cloudiness, and precipitation are required for model development and verification.

5.3 THE EFFECT OF OTHER GASES OF MAN-MADE ORIGIN

(a) Sulphur dioxide

The combustion of fossil fuels is the major source of man-made SO_2. The climatic effects of this gas could be considered to be the same as those of particles, as discussed above, since the SO_2 in the atmosphere is transformed to sulphate aerosol. Georgii (1979) has reviewed the large-scale distributions of SO_2 and its oxidation product, the sulphate aerosol, and discussed their climatic effects. He points out that, if sulphate from sea spray is ignored, 60% of the atmospheric sulphur is anthropogenic. This is illustrated in Figure 5.3, which shows the global sulphur cycle and emphasizes the considerable perturbation made to the global natural sulphur cycle by fossil-fuel combustion. In fact, the perturbation is even higher in regionally polluted areas, such as the eastern US, Canada, and Western Europe.

Georgii (1979) states that about 20% to 50% of the SO_2 emitted as gas into the atmosphere is transformed in the atmosphere to sulphate particles. The average tropospheric residence time is 1–3 days, while that of sulphate aerosols is 3–8 days. The sulphate can thus be observed up to 2,000–5,000 km from its source.

As a consequence of these observations, the climatic effect of sulphate aerosols, as opposed to SO_2, should be considered. The sulphate particles in non-cloudy areas cause scattering and absorption of incoming radiation, as well as absorption of outgoing radiation. Bolin and Charlson (1976) computed the effect on the scattering of solar radiation by sulphate particles and suggested that the change in solar radiation scattering in the polluted areas of eastern North America and Western Europe could correspond to a temperature change of several degrees. But they pointed out that in reality the effect is diluted because of the ventilation effect of the general atmospheric circulation. Distributing the effect over the whole

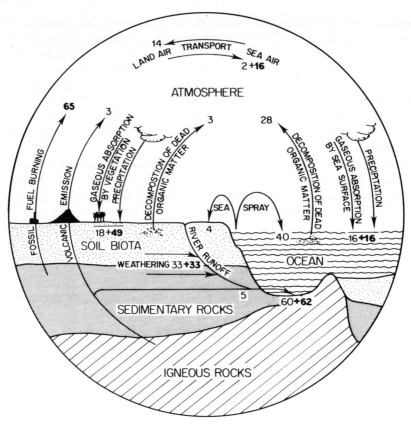

Figure 5.3 The sulphur cycle. The transfers are in 10^{12} g per year. Figures in lighter type show the transfers believed to have prevailed before man's activities significantly influenced the sulphur cycle. Bold face figures give estimates of additions due to man's activities. Source: Bolin and Charlson (1976)

Northern Hemisphere, Bolin and Charlson concluded that the scattering due to sulphate aerosols is equivalent to a drop of the average hemispheric temperature by 0.03–0.06 °C. However, they did not consider changes in infrared radiation or changes in cloud and precipitation processes. Moreover, Georgii (1979) has pointed out that, whereas the effect of sulphate on visible light is mainly that of scattering, this may not be true for some sulphate-containing aerosols, which also absorb light in the visible range. Therefore, as in the case of particles, the effect of increased sulphate levels in the atmosphere requires more detailed study.

(b) Oxides of nitrogen

In addition to interfering with the natural sulphur and carbon cycles, human activities also affect the nitrogen cycle. A detailed review of the natural sources and sinks of nitrogen and man-made perturbations of these is given by Hahn (1979), who states that major anthropogenic effects accrue from agriculture, forestry, combustion processes, and the breakdown of wastes. Combustion processes form oxides of nitrogen, ammonia, and nitrogen-containing pyrolysis products, and release them mostly into the atmosphere. The major part of this fixed nitrogen goes through nitrification/denitrification and is ultimately returned to the atmosphere in the form of dinitrogen oxide (N_2O) and molecular nitrogen.

Hahn (1979) suggests that large-scale environmental problems will occur if predictions about fertilizer use and combustion of fossil fuels are true. These problems include a depletion of stratospheric ozone due to the catalytic effect of man-made nitrogen oxide (NO) on the ozone destruction by atomic oxygen, and an increase of temperature of the atmosphere at the earth's surface due to the greenhouse effect of N_2O, NH_3, and HNO_3. Wang et al. (1976), using a one-dimensional radiative–convective model, calculated that a doubling of the present-day mixing ratios of N_2O, NH_3, and HNO_3 would lead to surface-temperature increases of 0.44 K, 0.09 K, and 0.06 K, respectively.

Hahn (1979) has examined the implications of these calculations for future growth in fertilizer use and combustion of fossil fuels. Figure 5.4 shows the results of the calculations for three sets of assumptions. Curve 2 represents mean values and suggests that a maximum warming of 1 K could occur due to N_2O. If it is assumed that NH_3 and HNO_3 have a behaviour similar to N_2O, then an additional 0.1–0.5 K from these gases could be considered. Thus, Hahn (1979) concludes that by the year 2000 the greenhouse effect due to increasing atmospheric N_2O may be about 40–50% of that due to increasing CO_2, while by 2050 it may be about 20–30%. However, Hahn (1979) points out that the factor that will determine the maximum tolerable level of atmospheric N_2O is probably not the greenhouse effect but rather the expected depletion of stratospheric ozone.

Isaksen (1980) has looked at the effect of nitrogen fertilization on the atmospheric N_2O content. He finds that there are large uncertainties attached to the production processes and concludes that the presently observed slow production rates suggest that no noticeable effect on climate or stratospheric chemistry will occur before late in the next century.

Recently, Donner and Ramanathan (1980) have investigated the effects of methane (CH_4) and N_2O on climate. They found that removing the present N_2O in the atmosphere would cool the earth–troposphere system by 1–2 W m^{-2}, while doubling the N_2O concentration would warm the system by 0.4–0.8 W m^{-2} (more warming in lower latitudes). Removing N_2O in the stratosphere would lead to cooling in the lower stratosphere and warming aloft. Donner and Ramanathan (1980) calculate that the hemispherical mean temperature increase for a doubling of the N_2O and CH_4 concentrations would be 0.3 K in each case. Wang et al. (1976) estimate 0.2 K and 0.44 K, respectively, for doubled CH_4 and doubled N_2O. The differences in the two sets of results are due to different sets of absorption data. Also the Wang et al. study used a one-dimensional model, while Donner and Ramanathan used latitudinal and seasonal calculations to obtain hemispherical means.

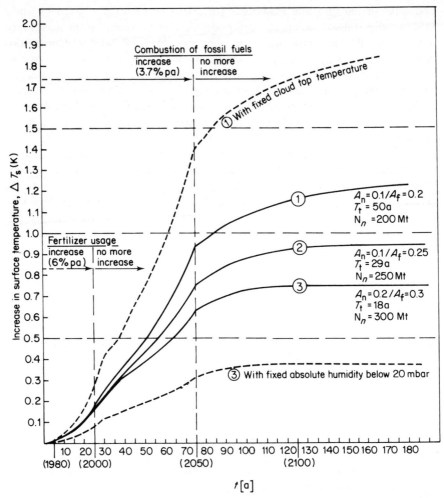

Figure 5.4   Increase in mean surface temperature due to increasing use of industrial fertilizers and fossil fuels. A_n and A_f are the yield factors for natural and agricultural soils, respectively; T_t is the tropospheric turnover time of N_2O; and N_n is the amount of nitrogen fixed by the biosphere each year. The dashed lines show the range of uncertainty of the climatic effect. Source: Hahn (1979)

(c) Acid precipitation

A particular environmental problem resulting from the release of oxides of sulphur and nitrogen into the atmosphere by the burning of fossil fuels has received increasing attention in recent years. It arises because these chemicals are returned to the earth as sulphuric and nitric acids in rain, snow, sleet, or fog. A recent report from NOAA (1981) has provided an outline of the problem and an extensive list of relevant literature. The following outline of the issue is based on this NOAA report.

Data gathered in recent decades in Canada, Japan, Scandinavia, the United States of America, and Western Europe have linked acid precipitation with possible damage to crops and forests, pollution of soils and water, killing of fish and other aquatic organisms, corrosion of materials, and climatic changes. The taller stacks constructed at power stations, refineries, and smelters have allowed the discharges to be carried further, and this has converted what was once a local concern into a regional or global one.

According to NOAA (1981), acid precipitation was first described in the 1960s by Norwegian scientists who found that the acidity of freshwater lakes had increased and fish catches had decreased. This was attributed to chemicals carried in the air from Western Europe. Similar changes in fish populations were noted in the northeastern United States in the 1960s and 1970s, but the recognition of the role of acid precipitation came only in 1970.

It is estimated that more than 33 million tons of oxides of sulphur and nitrogen are discharged annually into the atmosphere in the industrialized midwestern United States of America. The resulting acid precipitation is mainly recorded in the mountainous areas of New England, New York, and Pennsylvania.

The problem of acid precipitation has been recognized as a major, worldwide environmental problem. Major research programmes are underway in, for example, Canada, Norway, and Sweden. For detailed references to current research, the atmospheric chemistry and transport, aquatic effects, and the effects on soils and vegetation, the reader is referred to NOAA (1981).

(d) Other atmospheric trace gases

Carbon dioxide and the nitrogen compounds discussed above are not the only atmospheric trace gases that have a greenhouse effect. Wang et al. (1976) have pointed out that there is in fact a large number of trace gases which absorb long-wave radiation within the wavelength band in which most of the long-wave radiation is emitted from the earth's surface. These gases include N_2O, CH_4, NH_3, C_2H_4, SO_2, CCl_2F_2, CCl_3F, CH_3Cl, CCl_4. As indicated above, through the production and use of chemical fertilizers and the combustion of fossil fuels, the concentrations of some of these gases are being changed. Other gases (for example, the chlorofluorocarbons) have a purely anthropogenic origin. It is difficult to forecast the changes in the concentrations of these trace gases, but it is of interest to note the possible order of magnitude of temperature changes that could occur.

Wang et al. (1976) have used a radiative–convective model to compute the global average surface-temperature increase due to the greenhouse effect associated with increases in these trace gases. Table 5.2 shows the results of these computations. These results suggest increases in these atmospheric trace gases could have a climatic effect, especially in the case of N_2O, CH_4, and NH_3, each of

Table 5.2 Changes in global average surface temperature due to specified changes in atmospheric trace constituents. Results from radiative–convective model of Wang *et al.* (1976)

Trace gas	Assumed present concentration (ppmv)	Factor modifying concentration	ΔT^a (K) CCT[b]	ΔT^a (K) CTA[c]
N_2O	0.28	2	0.68	0.44
CH_4	1.6	2	0.28	0.20
NH_3	6×10^{-3}	2	0.12	0.09
HNO_3	*	2	0.08	0.06
C_2H_4	2×10^{-4}	2	0.01	0.01
SO_2	2×10^{-3}	2	0.03	0.02
CCl_2F_2	1×10^{-4}	20 ⎫	0.54	0.36
CCl_3F	1×10^{-4}	20 ⎭		
CH_3Cl	5×10^{-4}	2 ⎫	0.02	0.01
CCl_4	1×10^{-4}	2 ⎭		
H_2O	*	2	1.03	0.65
CO_2	330	1.25	0.79	0.53
O_3	*	0.75	−0.47	−0.34

a. Change in global average surface temperature.
b. Computed with the assumption of constant cloud-top temperature.
c. Computed with the assumption of constant cloud-top height.
*See comments by Wang *et al.* (1976).

which is influenced by the use of chemical fertilizers and the combustion of fossil fuels.

Chlorofluorocarbons, used in spray cans and refrigerants, have been a subject of concern in recent years because they are extremely stable non toxic man-made gases that persist in the troposphere for a long time. As Table 5.2 shows, a substantial increase in their atmospheric concentration would lead to a surface temperature increase, as Ramanathan (1975) has pointed out. If chlorofluorocarbons were produced at the 1973 production rates, then a temperature increase of 0.5 K by the year 2000 could occur. If the production rate continued to increase at 10% per year, as it was doing until recently, then a temperature increase of 1 K by 2000 could occur. A second effect of these gases arises because the molecules that diffuse up into the stratosphere are broken down by ultraviolet radiation and the resulting products could interfere with the ozone layer at that altitude (e.g., Crutzen, 1974; NAS, 1976).

While changes in the atmospheric concentration of chlorofluorocarbons and consequent changes in surface temperature and/or the ozone layer are not directly attributable to energy-conversion processes, the subject is mentioned here as an example of what Schneider (1977) has termed an 'energy externality', i.e., not energy-producing but nevertheless energy-dependent activities.

Wang *et al.* (1980) have recently extended their radiative–convective model studies of the effects of chlorofluoromethanes to look at the combined effects of the greenhouse effect and the effect of ozone depletion in the stratosphere on the surface temperature. Their study concentrates on the halogenated compounds $CFCl_3$, CF_2Cl_2, and CF_4, which they state are currently present in the atmosphere with mixing ratios of order 10^{-10} with potential for growth to concentrations $>10^{-9}$ due to continued anthropogenic release. It has been found that chlorofluoromethane-induced ozone depletion in the stratosphere can lower local temperatures by as much as 10 K (Ramanathan et al., 1976; Chandra et al., 1978). However, Wang et al. (1980) find that the effect on surface temperature due to depletion of stratospheric ozone is much smaller. To examine the effect of increases in chlorofluoromethanes, Wang et al. assume that $CFCl_3$ and CF_2Cl_2 are continually released at 1973 rates. In the year 2030 the atmosphere would then contain ∼0.8 and 2.3 ppbv $CFCl_3$ and CF_2Cl_2, respectively, compared with 0.1 and 0.2 ppbv at the present day. The greenhouse effect associated with the concentrations in the year 2030 is calculated to be 0.32 K. On the other hand, they calculate that a cooling of the surface temperature by ∼0.2 K due to ozone depletion would occur. They point out that their results depend critically on the detailed photochemistry of the lower stratosphere, which is subject to a range of uncertainties. The region of the upper troposphere and lower stratosphere is a complicated region to model and is most important for tropospheric climate studies. Wang et al. point out that further studies should address the problem with coupled dynamics, radiation, and photochemistry in this region.

The climatic effects of trace gases have also been discussed by Ramanathan (1980), who points out that anthropogenic sources of trace gases other than CO_2 (e.g., CH_4, N_2O, chlorofluoromethanes, and O_3) may contribute as much as 40% of the surface warming to the combined surface warming effects of CO_2 and these gases. Ramanathan finds that about 15% of the total warming is due to the projected increase in tropospheric ozone (O_3) caused by fossil-fuel sources of CO, NO, and CH_4. According to him an appreciable fraction of observed O_3 is produced within the troposphere by photochemical oxidation of CO, CH_4, and NO, which is roughly 25% (or more) of the natural source. Model results estimate that a continuation of the present growth rate of emission of these gases into the next century could lead to doubling the tropospheric ozone concentration. This could warm the climate by about 0.9 K, according to studies cited by Ramanathan. He points out that there is much uncertainty regarding the effect of trace gases, especially because about 25% of the warming effect is directly due to anthropogenic emission of trace gases, while the remaining 75% is a result of anthropogenic emission of photochemically active species which, through a complex series of chemical reactions, lead to an increase in radiatively active trace gases. He also emphasizes that tropospheric chemistry is a relatively new field and that the natural tropospheric sources and sinks of several of the gases are not known to within perhaps a factor of 2 or more.

One recent study has examined the sensitivity of a 12-layer tropospheric/ stratospheric general circulation model (GCM) to the addition of 10 parts per thousand million of chlorofluoromethanes (Dickinson and Chervin, 1979). The radiative effects of the chlorofluoromethanes were considered alone in one model run. In a second run, the ocean surface temperatures in the model were

158

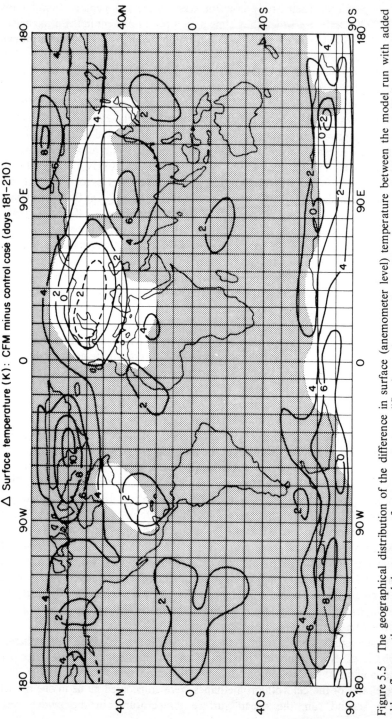

Figure 5.5 The geographical distribution of the difference in surface (anemometer level) temperature between the model run with added chlorofluoromethanes and the ocean surface temperature modification and a control case. The temperatures are averaged over days 181–210 of each model run. The stippled areas indicate statistically significant differences at the 10% significance level. Source: Dickinson and Chervin (1979)

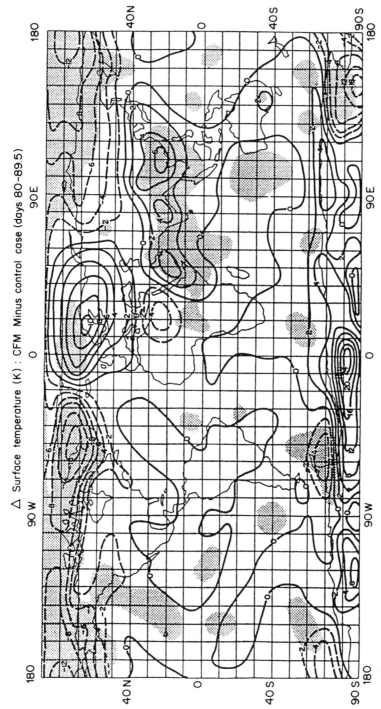

△ Surface temperature (K) : CFM Minus control case (days 80–89.5)

Figure 5.6   As in Figure 5.5 but for the model run with added chlorofluoromethanes only. Source: Dickinson and Chervin (1979)

simultaneously increased in order to simulate the effect of the changed radiative balance on ocean surface temperatures. Figure 5.5 shows the geographical distribution of the temperature differences at the surface between the control case and the case with chlorofluoromethanes and the ocean surface temperature increase. The shaded areas show where the differences are statistically significant according to calculations made by Dickinson and Chervin (1979). The response over the oceans is largely controlled by the prescribed ocean temperature changes of 2–4 K and therefore does not vary longitudinally to any large extent. Figure 5.6 shows the geographical distribution of the temperature differences at the surface between the control case and the case with only chlorofluoromethanes and no ocean surface temperature increase. We see that this pattern and the one shown in Figure 5.5 are out of phase in the higher latitudes of the Northern Hemisphere. Further detailed examination of other meteorological variables leads Dickinson and Chervin (1979) to the conclusion that the response patterns in atmospheric temperature were drastically different in the cases with and without ocean temperature changes. Several physical arguments to explain this are offered. The study shows a significant model response to a change in the global radiative balance of the imposed magnitude. The authors point out, however, that the specific practical information provided by the study regarding the detailed changes that would actually occur with increases of chlorofluoromethanes is minimal. This is so in particular because the calculated differences appear to be sensitive to whatever ocean temperature changes occur. The assumed ocean temperature change is estimated by Dickinson and Chervin (1979) to be correct within a factor of 2 of the longitudinal average of ocean changes that would occur and probably more uncertain in high latitudes. Since, in reality, the internal dynamics of the ocean would probably generate regional temperature variations just as large as the atmospheric differences, there is a need to investigate this problem using fully coupled ocean–atmosphere models when they are available.

(e) Concluding remarks

It is clear that, in addition to CO_2, man is adding other gases to the atmosphere that can have climatic effects. These effects have generally been investigated using one-dimensional radiative–convective models, although some preliminary sensitivity tests have been made with a global circulation model of the atmosphere. The effect of some gases, particularly N_2O and chlorofluoromethanes, on the ozone in the upper atmosphere, with consequent radiative and dynamical consequences for the lower atmosphere, indicates that models for studies on effects must be able to simulate the coupling of the troposphere and stratosphere. For a complete study of the effects of man's activities on climate, all of the trace gases will have to be considered together with complex models of the climate system.

REFERENCES

Atwater, M. A. (1975). Thermal changes induced by urbanization and pollutants. *J. Appl. Meteor.*, **14**, 1061–1071.

Auliciems, A., and I. Burton (1973). Trends in smoke concentrations before and after the clean air act of 1956. *Atm. Environment*, **7**, 1063–1070.

Bach, W. (1979). Impacts of fossil fuel use on environment and climate. *Proc. International DGLR/DFVLR Symposium on Hydrogen in Air Transportation*. Stuttgart, 11–14 Sept. 1979.

Bolin, B., and R. J. Charlson (1976). On the role of the tropospheric sulphur cycle in the shortwave radiative climate of the earth. *Ambio*, **5**, 47–54.

Bryson, R. A., and G. J. Dittberner (1976). A non-equilibrium model of hemispheric mean surface temperature. *J. Atmos. Sci.*, **33**, 2094–2106.

Chandra, S., D. M. Butler, and R. S. Stolarski (1978). Effect of temperature coupling on ozone depletion prediction. *Geophys. Res. Lett.*, **5**, 199–202.

Charlock, T. P., and W. D. Sellers (1980a). Aerosol, cloud reflectivity and climate. *J. Atmos. Sci.*, **37**, 1136–1137.

Charlock, T. P., and W. D. Sellers (1980b). Aerosol effects on climate: Calculations with time-dependent and steady-state radiative–convective models. *J. Atmos. Sci.*, **37**, 1327–1341.

Coakley, J. A., Jr., and P. Chylek (1975). The two-stream approximation in radiative transfer: Including the angle of the incident radiation. *J. Atmos. Sci.*, **32**, 409–418.

Crutzen, P. J. (1974). Estimates of possible future ozone reductions from continued use of chlorofluoromethanes (CF_2Cl_2, $CFCl_3$). *Geophys. Res. Lett.*, **3**, 169–172.

Dickinson, R. E., and R. M. Chervin (1979). Sensitivity of a general circulation model to changes in infrared cooling due to chlorofluoromethanes with and without prescribed zonal ocean surface temperature change. *J. Atmos. Sci.*, **36**, 2304–2319.

Dittberner, G. J. (1978). Climatic change; volcanoes, man-made pollution, and carbon dioxide. *IEEE Trans. on Geoscience Electronics*, GE-16, 50–61.

Donner, L., and V. Ramanathan (1980). Methane and nitrous oxide: Their effects on terrestrial climate. *J. Atmos. Sci.*, **37**, 119–124.

Eiden, R. (1979). The influence of trace substances on the atmospheric energy budget. In, W. Bach *et al.* (eds.), *Man's Impact on Climate*. Elsevier, Amsterdam.

Ellsaesser, H. W. (1975). The upward trend in airborne particulates that isn't. In, S. F. Singer (ed.), *The Changing Global Environment*, Reidel, Dordrecht, Holland.

Georgii, H.-W. (1979). Large-scale distribution of gaseous and particulate sulfur compounds and its impact on climate. In, W. Bach *et al.* (eds.), *Man's Impact on Climate*. Elsevier, Amsterdam.

Grassl, H. (1979). Possible changes of planetary albedo due to aerosol particles. In, W. Bach *et al.* (eds.), *Man's Impact on Climate*, Elsevier, Amsterdam.

Hahn, J. (1979). Man-made perturbation of the nitrogen cycle and its possible impact on climate. In, W. Bach *et al.* (eds.), *Man's Impact on Climate*. Elsevier, Amsterdam.

Isaksen, I. S. A. (1980). The impact of nitrogen fertilization. In, W. Bach, J. Pankrath, and J. Williams (eds.), *Interactions of Energy and Climate*. Reidel, Dordrecht, Holland.

Kellogg, W. W. (1977). Effects of human activities on global climate. Techn. Note No. 156, WMO No. 486, World Meteorological Organization, Geneva.

Kellogg, W. W. (1980). Aerosols and climate. In, W. Bach, J. Pankrath, and J. Williams (eds.), *Interactions of Energy and Climate*. Reidel, Dordrecht, Holland.

Landsberg, H. E. (1975). Man-made climatic changes. In, S. F. Singer (ed.), *The Changing Global Environment*. Reidel, Dordrecht, Holland, pp. 197–234.

Landsberg, H. E. (1979). The effects of man's activities on climate. In, M. R. Biswas and A. K. Biswas (eds.), *Food, Climate, and Man*. John Wiley and Sons, New York.

Liou, K.-N. (1976). On the absorption, reflection, and transmission of solar radiation in cloudy atmospheres. *J. Atmos. Sci.*, **33**, 798–805.

Manabe, S., and R. T. Wetherald (1967). Thermal equilibrium of the atmosphere with a given distribution of relative humidity. *J. Atmos. Sci.*, **24**, 241–259.

Mitchell, J. M., Jr. (1975). A reassessment of atmospheric pollution as a cause of long-term changes of global temperature. In, S. F. Singer (ed.), *The Changing Global Environment*. Reidel, Dordrecht, Holland, pp. 149–173.

162

Munn, R. E., and L. Machta (1979). Human activities that affect climate. *Proc. of the World Climate Conference.* WMO Publication No. 537, World Meteorological Organization, Geneva.

Naegele, P. S., and W. D. Sellers (1981). A study of visibility in eighteen cities in the western and southwestern United States. *Mon. Wea. Rev.*, **109**, 2394–2400.

NAS (1976). Halocarbons: Environmental effects of chlorofluoromethane release. Committee on Impacts of Stratospheric Change, National Academy of Science, Washington, DC.

NOAA (1981). *Acid Precipitation.* Current Issue Outline 81-1, National Oceanic and Atmospheric Administration, Rockville, Maryland, USA.

Pollack, J. B. *et al.* (1976). Estimates of climatic impact of aerosols produced by space shuttles, SST's and other high flying aircraft. *J. Appl. Meteor.*, **15**, 247–258.

Ramanathan, V. (1975). Greenhouse effect due to chlorofluorocarbons: Climatic implications. *Science*, **190**, 150–152.

Ramanathan, V. (1980). Climatic effects of anthropogenic trace gases. In, W. Bach, J. Pankrath, and J. Williams (eds.), *Interactions of Energy and Climate.* Reidel, Dordrecht, Holland.

Ramanathan, V., L. B. Callis, and R. E. Boughnev (1976). Sensitivity of surface temperature to perturbations in the stratospheric concentration of ozone and nitrogen dioxide. *J. Atmos. Sci.*, **33**, 1092–1112.

Rasool, S. I., and S. H. Schneider (1971). Atmospheric carbon dioxide and aerosols: Effects of large increases on global climate. *Science*, **173**, 138–141.

Reck, R. A. (1974). Aerosols in the atmosphere: Calculation of the critical absorption/backscatter ratio. *Science*, **186**, 1034–1035.

Reck, R. A. (1975). Aerosols and polar temperature changes. *Science*, **188**, 728–730.

Robinson, G. D. (1977). Effluents of energy production: Particulates. In, *Energy and Climate.* National Academy of Sciences, Washington, DC.

Russell, P. B., J. M. Livingston, and E. E. Uthe (1979). Aerosol-induced albedo change: Measurement and modeling of an incident. *J. Atmos. Sci.*, **36**, 1587–1608.

Schaefer, V. J. (1975). The inadvertent modification of the atmosphere by air pollution. In, S. F. Singer (ed.), *The Changing Global Environment.* Reidel, Dordrecht, Holland.

Schneider, S. H. (1977). What climatologists can say to planners. In, *Living with Climatic Change, Phase II.* The Mitre Corporation, McLean, Virginia.

Toon, O., and J. Pollack (1976). A global average model of atmospheric aerosols for radiative transfer calculations. *J. Appl. Meteor.*, **15**, 225–246.

Twomey, S. (1972). The effect of cloud scattering on the absorption of solar radiation by atmospheric dust. *J. Atmos. Sci.*, **29**, 1156–1159.

Twomey, S. (1977). The influence of pollution on the shortwave albedo of clouds. *J. Atmos. Sci.*, **34**, 1149–1152.

Viskanta, R., R. W. Bergstrom, and R. O. Johnson (1977). Radiative transfer in a polluted urban planetary boundary layer. *J. Atmos. Sci.*, **34**, 1091–1103.

Wang, W., and G. M. Domoto (1974). The radiative effect of aerosols in the earth's atmosphere. *J. Appl. Meteor.*, **13**, 521–534.

Wang, W.-C., J. P. Pinto, and Y. L. Yung (1980). Climatic effects due to halogenated compounds in the earth's atmosphere. *J. Atmos. Sci.*, **37**, 333–338.

Wang, W.-C., Y. L. Yung, A. A. Lacis, T. Mo, and J. E. Hansen (1976). Greenhouse effects due to man-made perturbations in trace gases. *Science*, **194**, 685–690.

Weare, B. C., R. L. Temkin, and F. M. Snell (1974). Aerosol and climate: Some further considerations. *Science*, **186**, 827–828.

CHAPTER 6

The Effect of Solar Energy Systems on Climate

6.1 INTRODUCTION

As the IIASA Energy Systems Program (1981) has discussed in more detail, there are several renewable energy technologies that together could provide an essentially inexhaustible, long-term energy supply on a global scale: solar energy, wind, hydropower, biomass, geothermal energy, etc. It is common to distinguish between 'hard' and 'soft' technologies, where 'soft' refers to simple technologies, such as harvesting of wood from forests and using small-scale hydropower. The term 'decentralized' is used to describe localized systems not part of centralized supply systems. It is expected that future energy systems will employ a mixture of renewable energy technologies. It is clear that the effects on climate could be on the local—or possibly global—scale, depending on the application.

The solar technologies considered by the IIASA Energy Systems Program (1981) within the context of 'hard' solar energy use are: solar thermal-electric conversion, in which direct solar radiation is concentrated on to an absorber and the high temperatures thus produced are used to drive turbines and generate electricity; photovoltaic electricity, in which diffuse and direct solar radiation is converted by photovoltaic cells; biotechnology, involving photosynthetic and biological energy conversion; and the solar satellite power station, in which solar cells are used to capture solar energy in space and the energy is radiated in microwave form to a collector on the earth's surface.

The Energy Systems Program (1981) suggests that a 'hard' or global solar energy system would have the following features:

- Local use of solar-generated heat for space heating, water heating, and industrial process heat where economically and logistically suitable.
- Local and regional use of small-scale, solar-mechanical, solar-electrical, and solar-fuel generating units, especially in developing countries.
- Solar electric power plants of various sizes located throughout the world, primarily in sunny regions, interconnected through large integrated electric utility systems over distances of many thousands of kilometres.
- Solar fuel generation units primarily in sunny regions interconnected globally via pipeline or tanker.

163

Table 6.1 Realizable potential of individual renewable energy sources. Source: Energy Systems Program (1981)

Source	Application	Quantity (TWyr/yr)	Energy form
Biomass	In poor rural areas	0.8	Charcoal
		0.1	Biogas
	Other farm waste	0.1	Biogas
	Forest products industries	0.7	Coal equivalent
		0.1	Electricity
	Energy plantations	3.0	Primary energy to be processed
	Urban waste	0.3	Coal equivalent
Hydroelectricity		1.5	Electricity
Wind		1.0	Electricity
Direct solar heat		0.9	Low temperature heat
Geothermal (wet)		0.5	Low temperature heat
		0.1	Electricity
Other renewable sources (mostly OTEC)		0.5	Electricity
TOTAL		9.6	

In contrast to the large-scale global uses of solar energy, there are other 'soft' or smaller-scale, decentralized applications. Table 6.1 shows estimates of the realizable potential of these solar energy sources and another renewable source, geothermal, which will not be considered further in this chapter. The potential of the renewables is found by the Energy Systems Program (1981) to be equivalent to that from 14–15 TWyr/yr of fossil fuels or nuclear power, assuming that one must make an equivalent mix of secondary energy carriers. This suggests that soft uses of renewable energy sources have potentially the same order of magnitude as fossil fuels or nuclear power in future energy strategies.

This chapter will discuss the effects on climate of solar energy technologies. The effects on climate of solar energy conversion systems have not received as much attention in the past as those of fossil fuel and nuclear systems. However, a workshop was held at IIASA (Williams et al., 1977), which made a preliminary evaluation of the available systems, their physical characteristics, potential perturbations to climatic boundary conditions, and the climatic implications.

6.2 THE POTENTIAL INFLUENCE OF SOLAR TECHNOLOGIES ON LOCAL AND REGIONAL CLIMATE

(a) Solar thermal electric conversion

Systems that use turbines to convert solar energy into electricity have received considerable attention in recent years and systems with capacities ranging from

1 kW(e) to 5 MW(th) exist. Some of the small-scale systems use flat-plate solar energy collectors and an organic working fluid that drives a Rankine cycle engine, but these systems have low efficiencies of only about 3% because of the low working temperatures.

For better efficiencies higher temperatures are required. One alternative is to use concentrating collectors (e.g., trough or parabolic collectors) and then transport the fluid to the turbine. In this case, efficiencies of 15–20% can be obtained, but only direct radiation can be concentrated. A second possibility is the central receiver or 'power tower' design, which has been evaluated as the economically preferable system for large-scale units. The system consists of a central tower with a receiver on the top and several thousand steered mirrors (heliostats) on the surrounding ground, which reflect direct solar radiation on to the receiver. So far the largest installation of this type (10 MW(e)) is located in California. In a location favoured by a high proportion of direct radiation (70–80% of global radiation), a central receiver plant employing a conventional steam turbine may be operated with an overall efficiency of 15–18%.

Figure 6.1 illustrates the design of a 100 MW(e) solar thermal electric conversion (STEC) power plant of Boeing Engineering and Construction (EPRI, 1976). Certain characteristics are noteworthy, especially with regard to their potential influence on climate. The central receiver is 260 m high and is surrounded by

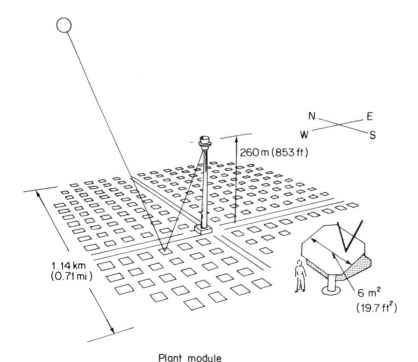

Plant module

Figure 6.1   Design of a 100 MW(e) central-receiver-type solar thermal power plant. Source: EPRI (1976)

heliostats over an area of 1.3 km². The heliostats (total of 15,400) cover about 39% of the total ground surface and each collector has an area of about 32 m².

Because of the dependence on favourable insolation, central receiver systems are more likely to be situated in dry, cloudless areas, e.g., the desert areas of North Africa and the southwest USA. However, the potential for large-scale deployment of solar thermal electricity generation in southern Europe has also been considered (Jäger et al., 1978). Weingart (1978) suggests that production of solar thermal electricity in sunny regions and shipment (by long-distance, high-voltage transmission lines) to less sunny areas would be far cheaper than generation in the latter.

The major changes in boundary conditions of the climate system due to installation of a STEC system would be:

- changes in the surface energy balance;
- changes in surface roughness, since heliostats are up to 10 m high;
- changes in surface hydrological characteristics if the area is paved.

The effect of a STEC system on the surface energy balance would in reality be extremely complex, since so many variables are perturbed. For example, the changes in surface thermal characteristics through paving, the changes in energy fluxes from the surface through the roughness changes, and other micro-meteorological changes must be considered. A simplified estimate of the effect of STEC systems on the energy balance can be made, however, by considering the basic components of the energy balance.

Figure 6.2 shows a simplified description of the energy balance in the absence of a STEC plant. Of the direct insolation, 30% is reflected away from the surface, 70% is absorbed and then is re-emitted in the form of long-wave radiation, sensible heat, and latent heat. As a rough estimate Figure 6.2 divides the energy

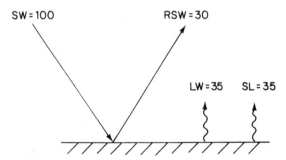

Figure 6.2 Simplified illustration of the energy balance at the earth's surface in the absence of a STEC plant. SW = incoming solar radiation, RSW = reflected solar radiation, LW = long-wave radiation, SL = sensible plus latent heat flux. Source: Williams and Krömer (1979)

flux from the surface equally between the long-wave radiation and sensible plus latent heat.

In the presence of a STEC plant (Figure 6.3) the energy balance is changed differently according to the season. Assuming a STEC plant with a ground cover ratio of 40%, at which 17% of the sunlight impinging on the heliostats is converted to electricity, then the following estimates of the effect on the energy balance can be made. In winter, the sun is low in the sky, thus the heliostats intercept all of the insolation on the area, i.e., there is no reflection from the ground. The heliostats themselves have an efficiency of, say, 85%, that is, 15% of the light hitting the heliostats is lost due to aiming errors, haze, and optical-surface imperfections. In addition, 10% of the radiation hitting the heliostats will be absorbed by the mirrors. In winter, therefore, the reflection from the STEC area is 14% of the total incoming direct radiation, i.e.:

total incoming on mirrors   100%
absorbed by mirrors   10%
optical losses = 15% of the amount reflected = 15% × 90% = **14%**.

In summer the sun is high in the sky, so that only 40% of the ground is covered by

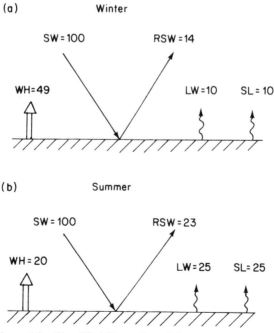

Figure 6.3   Simplified illustration of the energy balance at the earth's surface with a STEC plant (a) in winter, (b) in summer. W = waste heat. Other abbreviations as in Figure 6.2. Source: Williams and Krömer (1979)

heliostats, thus incoming radiation is also reflected by the ground. Reflection from the ground is then 60% × 30%, assuming that the ground has the same reflectivity as in the absence of the STEC plant. Of the radiation incident on the heliostats, 10% is again absorbed and of the remaining 90%, 15% is lost optically through aiming errors, etc. Thus, for the summer case, the reflection from the STEC area is 23% of the incoming direct radiation, i.e.:

reflection from ground   = 30% × 60% = **18%**
plus
total incoming on mirrors = 40%
absorbed by mirrors   = 4%
optical losses = 15% of the amount reflected = 15% × 36%
                                               = ~**5%**
             total reflection (summer)   = **23%**

In the winter season absorption by the soil is 0%, since the heliostats intercept all the incoming radiation, while in the summer, of the 60% of incoming radiation reaching the ground, 14% is reflected (see above) and, thus, 46% is absorbed. As indicated above, the mirrors themselves absorb 10% of the incident radiation.

Of the radiation reaching the receiver there are losses, which have been estimated as 6% of the incident radiation on the heliostats, i.e., 6% of the total incident radiation in winter and 2% (= 40% × 6%) of the total incident radiation in summer. In addition to the receiver losses, there are so-called 'piping' losses, which have been estimated as 4% of the incident radiation on the heliostats, i.e., 4% of the total incident radiation in winter and 2% (= 40% × 4%) of the total incident radiation in summer.

Over the STEC area as a whole, therefore, the energy leaving the surface (ground plus plant) is, expressed as a percentage of the total incident radiation on the whole area,

	winter	summer
from the soil (equals that absorbed)	0%	42%
from mirrors (equals that absorbed)	10%	4%
from receiver	6%	2%
from piping	4%	2%
	sum 20%	50%

Assuming, as done for the case without a STEC plant, that this energy flux from the surface is divided equally between long-wave radiation and sensible plus latent heating, then the above sum can be divided:

	winter	summer
long-wave radiation	10%	25%
sensible plus latent heating	10%	25%

Of the radiation incident upon the heliostats, 17% is converted to electricity, that is, in winter 17% of the total incident radiation on the STEC plant and in summer 7% (40% × 17%) of the total incident radiation. The remainder of total incident radiation is emitted as waste heat at cooling towers. The amount of waste heat, as a percentage of the total incident radiation is:

	winter	summer
reflected from the area	14%	23%
soil, mirror, receiver, piping losses	20%	50%
electricity generated	17%	7%
subtotal	51%	80%
remainder = waste heat at cooling towers	49%	20%

Figure 6.3 shows these numbers diagrammatically; a comparison with Figure 6.2 shows that in the presence of the STEC plant the total of the energy flux from the surface into the atmosphere (~70% of the incident solar radiation on the area) does not differ from that in the absence of the STEC plant. However, the distribution is changed. The long-wave radiation from the surface is reduced from the 35% in the case of no STEC plant to 10% (winter) or 25% (summer) in the presence of the STEC plant. Similarly, the sum of sensible and latent heat flux from the surface is reduced to 10% (winter) and 25% (summer). The significantly lower heat release from the surface is compensated by a release of waste heat from cooling towers upon energy conversion. In this respect some effects of STEC systems upon climate can be evaluated in the same way as the potential effect of waste heat from fossil fuel or nuclear power plants. As with the discussion of the effect of waste heat release (see Chapter 4), the effect of surface-energy-balance changes will depend on the scale (horizontal dimensions) and the magnitude of the perturbation.

The effect of such energy-balance changes has been investigated by Bhumralkar *et al.* (1979) using a two-dimensional mesoscale model of the atmosphere. For this study the plant characteristics were based on the central receiver STEC plant system of Martin Marietta (1975). The atmospheric effects of a hypothetical STEC facility located in southern Spain were investigated. To examine the effects of large-scale installations, an area of about 1,000 km² was assumed; such a plant could generate about 30 GW and might be considered as a solar power park. Figure 6.4 shows the location of the STEC plant. The horizontal dimension of the STEC plant was assumed to be 32 km (oriented in the direction of the prevailing wind), with heliostats covering 8 km (ground cover ratio 25%; Hildebrandt and Vant-Hull, 1977). The total horizontal dimension considered was 400 km, so that meteorological conditions upstream and downstream of the STEC plant could be included.

Figure 6.5 shows the land surface divisions used in the study, together with the components of the surface energy balance for each of the surface types. It was assumed that the STEC system uses wet cooling towers for dissipating waste heat

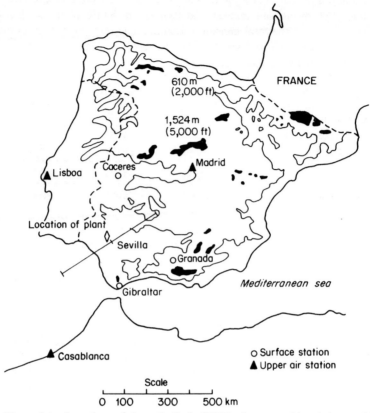

Figure 6.4   Location of hypothetical STEC plant considered in model
simulations. The line shows the length and direction of the horizontal domain
of the model. Source: Bhumralkar *et al.* (1979)

into the atmosphere and this heat is released at heights of 100 m or more in the
form of latent heat (80%) and sensible heat (20%).

The model was run for typical conditions of summer and winter in Spain, as
derived from meteorological observations. The integration was for a simulation of
9 hours, 8 a.m. to 5 p.m. Figure 6.6 shows the relative humidity for the initial conditions and after 5.5, 7, and 9 hours simulated time for the summer cases with and
without STEC. The power plant has caused the model atmosphere to be relatively
more moist than in the control case, with cloud development (where cloud is
defined as an area with relative humidity of more than 100%). In the control case,
clouds are sporadic and cover relatively small areas of domain, whereas they are
quite pronounced and persistent in the 'with STEC' case. These results are
supported by similar considerations of the temperature and vertical velocity fields.

The model was also integrated with initial conditions corresponding to winter
meteorological conditions in southern Spain. In the winter simulation, no clouds
formed anywhere in the model domain throughout the simulation in both the case

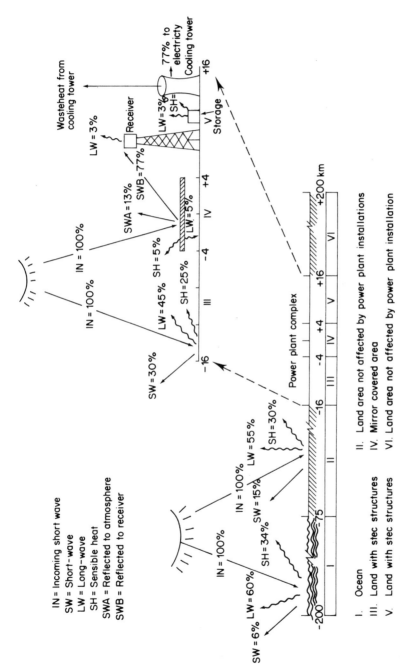

Figure 6.5 Types of land or ocean surface and their energy-balance characteristics used in model simulations. (SH includes the sensible and latent heat release to the atmosphere, considered to have a constant proportionality.) Source: Bhumralkar *et al.* (1979)

172

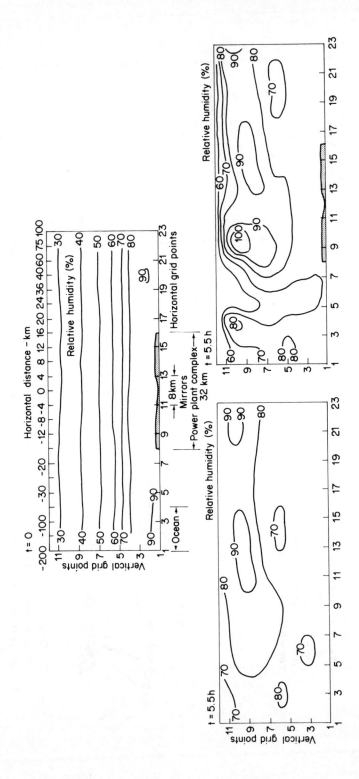

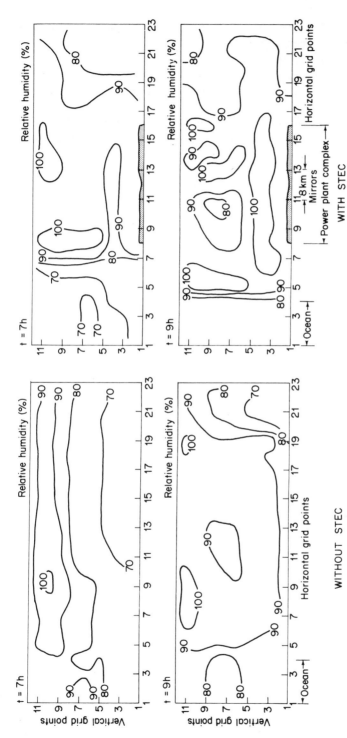

Figure 6.6 The horizontal–vertical distribution of relative humidity in the initial conditions (above), in summer simulation without STEC plant (left-hand side) and in summer simulation with STEC plant (right-hand side). The results of the simulations are shown for 5.5 hours (above), 7 hours (centre), and 9 hours (below) from the beginning. Source: Bhumralkar *et al.* (1979)

without STEC and that with STEC. It is suggested that this is due to the strong prevailing winds assumed for the winter simulation.

Bhumralkar *et al.* (1979) emphasize that the model results should be treated as preliminary. They suggest the need to investigate these effects for a wide range of meteorological conditions and with variations in the design, areal extent, capacity, and location of the STEC system. Bhumralkar (personal communication, 1979) has recently used the same model to study the effect of a 10 MW STEC plant being built in Barstow, California, and preliminary results indicate that the location plus the characteristics of the STEC facility are again sufficient to produce local meteorological effects.

The effects of the other potential changes due to STEC systems, i.e., those in roughness and wetness of the surface, have not been investigated in the same detail as the energy-balance changes. Some implications of roughness changes are discussed below with reference to wind-energy systems and some implications of wetness changes are described with reference to biomass systems.

(b) Photovoltaic conversion

Photovoltaic power systems consist of arrays of small individual generating units, the photovoltaic cells, in which semiconductor materials are used to convert the energy in light to electricity. The large interest in photovoltaic electricity generators is largely because they can be made in a wide range of sizes, from small units with a few watts output to the gigawatt designs of solar power satellites. A second advantage is that photovoltaic units can convert diffuse as well as direct solar radiation, so that they are attractive for the higher, more cloudy latitudes. An important point with regard to photovoltaic systems is that there are no apparent economies of scale (Kelly, 1978) and smaller systems have a number of advantages.

Photovoltaic cells are generally made from silicon, cadmium sulphide/copper sulphide, or gallium arsenide. For individual single-crystal silicon cells, efficiencies of 12–14% are found, but commercially available arrays have an efficiency of about 8% because of incomplete coverage of the array areas because of mismatches among cells (Kelly, 1978). The efficiencies of cadmium sulphide/copper sulphide cell arrays, which are suited to mass production, are not more than 4% (Arndt *et al.*, 1978). Gallium arsenide single-crystal cells have efficiencies close to 15%, but since these cells are expensive they are generally combined with concentrators to reduce the required cell areas.

There are a number of ways in which photovoltaic cells may be deployed, ranging from arrays of cells on the roofs of houses to modules of cell arrays for a central power station. At present, the use of photovoltaic conversion in a decentralized fashion appears to be the preferred option. In this case, the local/regional effects on weather and climate would presumably not differ in magnitude from the effects of urban areas. Isolated applications of decentralized photovoltaic systems cannot be expected to affect the local weather, a single 10 kW array on a rooftop would only exert a micrometeorological influence on the conditions in a narrow layer over the roof. On a larger scale, if photovoltaic systems were used on a high proportion of rooftops, over parking areas, etc., in an urban area, then their effect would essentially be an enhancement of the urban heat-island effect, which was discussed in Chapter 4.

As a rough estimate of the effect of a flat-plate collector system, one can assume that, owing to the use of non-reflecting surfaces, the cells absorb 95% of the radiation falling on them and that the cells cover most of the flat plate (again 95%). Therefore a very high percentage of the insolation of the flat plates is absorbed. However, cell efficiency is low, only 10% or so of the energy is converted to electricity and the rest of the energy must be released into the atmosphere as longwave radiation, sensible and latent heat. With cells of only 5–8% efficiency, more energy must be relased. Over natural surfaces the amount of energy that is absorbed and subsequently released to the atmosphere varies: 70–75% for desert areas, 75–80% for dried grass, 85–93% for coniferous forests, 90–93% for water (Munn and Machta, 1979). Thus, if the photovoltaic system were installed in an area where the surface was naturally dark (e.g., forest, wet fields), there would be no significant change in the amount of energy input to the atmosphere. At the other extreme, if cells with a low efficiency were used and the natural area had a high reflectivity (e.g., desert, dry grass), then the photovoltaic system would act as a heat source. However, for the EPRI (1978) 200 MW(e) flat-plate reference system, this heat source would have an area on the order of only 3 km^2. For this size of perturbation the percentage change in the amount of energy input would probably not be enough to cause significant effects on the local weather or climate. It should be noted that the photovoltaic systems would lead to an increased proportion of sensible heat input to the atmosphere and a decreased proportion of latent heat input.

The presently conceived photovoltaic conversion systems and the scale of their deployment, therefore, appear to have only a potential effect on micrometeorological conditions, on the scale of the collectors themselves. A significant effect on local/regional climate could only be foreseen in the case of massive deployment of the systems, say in desert areas, or installation of plants with thousands of MW(e) capacity. Since, however, there appear to be no economies of scale and since there are advantages in smaller-sized systems, this is unlikely.

For reference purposes only, as a guide to the potential effects of larger-scale changes in energy input to the atmosphere, a few studies can be cited. An early suggestion that weather can be modified by changing the reflectivity of the surface was made by Black (1963), who suggested that coating the ground with black asphalt could increase soil temperatures as much as 10 °F, whereas if the freshly applied asphalt were used to bind a white material such as lime, subsurface temperatures could be reduced as much as 13 °F. Black pointed out that surface heat sources play a significant role in localizing cumulonimbus clouds and thundershowers, when atmospheric conditions are suitable. Therefore albedo changes were suggested as possibly effective in modifying weather.

Charney et al. (1977) have carried out a number of model simulations to investigate the effects of albedo changes on the atmospheric circulation. As part of this study, they evaluated the least dimension of an area for which albedo and ground moisture changes can be expected to influence convective rainfall. It was suggested that a plausible rule of thumb is that observable effects can be expected when the characteristic time for a change in the moist static energy to penetrate to cloud base is smaller than the time required for new properties to be advected into the region. This suggests a minimum dimension of 40–80 km.

Berkofsky (1976) has developed a local atmospheric circulation model and has used it to show that the vertical circulation and therefore cloudiness and precipitation respond to surface albedo changes in desert areas. In particular, the model showed that a lowering of the surface albedo in a desert region could lead to increased vertical velocity and possibly increased rainfall.

(c) Biomass conversion

As suggested in Table 6.1 and by reviews such as those of Hayes (1977) and Hall (1979), there is a wide variety of biological energy sources, which can be roughly divided into waste from non-energy processes (e.g., food and paper production) and crops grown explicitly for their energy value. Biological sources of energy have several advantages. They can produce solid, liquid, and gaseous fuels. Conversion is possible on a local, decentralized scale using generally relatively simple technology and larger-scale applications are also possible. In developing countries, biomass has traditionally supplied most of the energy used. In developed countries substantial sources exist already in wastes and could be increased by growing plants specifically for energy conversion. In addition, as Hall (1979) and the Energy Systems Program (1981) have discussed, possible future non-conventional uses of photosynthesis for such processes as hydrogen production could be introduced.

The IIASA Energy Systems Program (1981) estimates the technical potential of biomass energy sources to be 6 TWyr/yr of secondary energy. However, owing to limitations such as competitive land use and economics, they consider that the realizable potential of biomass energy sources will be 5.1 TWyr/yr (see Table 6.1). This energy would be in several forms, including biogas, charcoal, wood to be burned or otherwise processed, and electricity from turbines driven by the heat from burning wood wastes in the pulp and paper industry.

The heterogeneity of biomass energy sources means that no simple evaluation of the local and regional effects of biomass conversion on climate can be given. It is likely that biomass conversion could indeed be deployed in a decentralized fashion and each locality would choose a conversion system most suited to the local sources of biomass supply and requirements for energy. In the study for a possible 'Solar Sweden' (Johansson and Steen, 1978), in which it is proposed that half of the energy required in the year 2015 would be from biomass, energy plantations of 3 million hectares would be required. However, the forestry industry in Sweden presently uses 23 million hectares and thus the effect of future energy conversion would be presumably smaller than that of present forests. Likewise, a very high proportion of the firewood cut today is burned for cooking food and warming houses. The biomass conversion schemes to use firewood in the future generally are based on technology to use firewood far more efficiently than it is used today, thus providing more energy from fewer resources.

The above points suggest that no comprehensive evaluation of the effect of biomass conversion on local weather and climate can be given. The effects will depend on the technology chosen, the scale on which it is used, and the location in which it is applied, and in certain cases such effects will be smaller than those due to human activities already undertaken. Nevertheless, a few general points can be made regarding the local influences on weather and climate of biomass conversion systems.

In view of the potential role of wood in biomass systems, the climate-related properties of forests should be considered. Baumgartner (1979) has given a review of the interactions between forests and climate on which this discussion is based. From the point of view of energetics, forests are the most active elements of the land cover. The energy exchange is not concentrated in a thin layer, because radiation is intercepted and transferred within the stand. Consequently, forests do not overheat greatly relative to the ambient temperature and the long-wave radiation emission is therefore low. But trees have a low reflectivity for short-wave radiation ($\sim$10% for coniferous and $\sim$15% for deciduous trees) and the net radiation over forests is high. Baumgartner (1979) gives the net radiation for forests as 67 J m^{-2}, for crops as 53 J m^{-2}, for grass as 47 J m^{-2}, and for bare soil as 33 J m^{-2}. The high-energy income is used by the forests primarily for evapotranspiration. The potential evaporation from forests (850 mm per annum) is twice that of bare soil (Baumgartner, 1979). Forests consume more water than other vegetation and transfer it by evapotranspiration back into the atmosphere. The increased water vapour in the air results in generally higher precipitation over forest areas. But, despite their use of water, forests also act as an important control of surface water. Moltschanov (1966), cited by Baumgartner (1979), showed for the southern USSR that in totally forested areas only 18% of the precipitation appears as river discharge, whereas in open land 42% appears as discharge.

The apportionment of net radiation into sensible and latent heat is usually discussed in terms of the Bowen ratio (i.e., the ratio of sensible heat transfer to latent heat transfer), which gives a good measure of the effect of forests on the energy balance. Above irrigated rice fields, swamps and tropical rainforests, the Bowen ratio is well above 1, with values of about 10 in desert regions (SMIC, 1971).

As Baumgartner has pointed out, forests also have a dynamic interaction with the atmosphere. The top of the forest is one of the roughest earth–atmosphere interfaces, with the crown braking the movement of the air and absorbing momentum. This effect can be expressed by the roughness parameter z_0, values of which, for various surfaces, are given in Table 6.2.

The effect of the increased 'roughness' of forest compared with open land is illustrated in Figure 6.7, which shows the influence on wind velocity profiles exerted by a dense 45-year-old ponderosa pine forest in California and an oak grove. According to Barry and Chorley (1968), measurements for European

Table 6.2   The roughness parameter z_0 for various surfaces. Source: Baumgartner (1979)

	Water	Sand	Short grass	Crops	Forest	City
z_0(cm)	0.01	0.2	1–5	10	300	400

forests show that a 30 m penetration reduces wind velocities 60–80% and the reduction is to 7% with a penetration of 120 m. Baumgartner (1979) emphasizes that forests influence, thereby, the surface drag, energy dissipation, and turbulent exchange of the air, which all depend on the roughness; thus any large-scale change in forest cover could have an effect on the development of the pressure field and the circulation of the atmosphere. It is further pointed out that the increased roughness in forests produces high turbulent exchange and mixing rates, which are very active in cleansing polluted air near cities or industries.

Another potential effect of biomass conversion is through the production of particulates by burning biomatter or through windblown dust from agricultural land. Several factors have to be taken into account when considering the role of particles in the atmosphere. Firstly, there are large uncertainties regarding present natural and anthropogenic production rates, as has been discussed in Chapter 5 (see Table 5.1). Secondly, larger particles (>5 μm) fall out of the atmosphere rapidly and even smaller particles in the lower atmosphere are soon washed out by precipitation processes. Thus, the average lifetime of a particle is of the order of 10 days, and most particles are found in the atmosphere close to their source. However, volcanic particles and secondary aerosols, if injected into the stratosphere, can exist there for a long time and become widely distributed. The effects of particles on climate and weather are various. As indicated in Chapter 5, particles have an influence on the condensation/precipitation process. The particles can also absorb radiation and warm the atmosphere or reflect radiation, giving a surface cooling; such effects are very dependent on the characteristics of the particles and of the underlying surface.

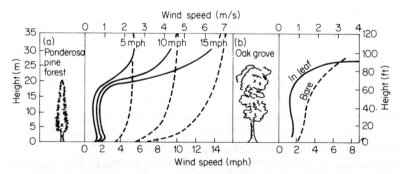

Figure 6.7   Influence on wind velocity profiles of (a) a dense 45-year-old ponderosa pine stand, the dashed lines show the wind profile over open country; and (b) an oak grove. Source: Barry and Chorley (1968)

Table 6.3   Estimates of particle production rates for particles with diameter less than 5 μm. Source: Dittberner (1978)

	10^6 tonnes yr^{-1}	Per cent
Anthropogenic input		
Fossil fuels		
Gases (subsequently converted to		
particles in the atmosphere)	311	45
Particulates	54	8
Wind-blown dust	180	26
Agricultural burning		
Hydrocarbons	72	10
Particulates	62	9
Nitrates	7	1
Fuel wood		
Particulates	4	1
Forest fires		
Particulates	2	0
	692	100
Natural input (excluding volcanoes)		
Sea salt	500	46
Sulphates (natural decay)	335	31
Wind-blown dust (natural decay)	120	11
Hydrocarbons (natural decay)	75	7
Nitrates (natural decay)	60	5
Forest firest (lightning)	3	0
	1093	100
Volcanic input		
Dust	25–150	37
Sulphates	42–255	63
	67–405	100
Grand Total	1852–2190	

Table 6.3 shows estimates of particulate production at the present day; the data are derived from Dittberner (1978) as given by Munn and Machta (1979). It is clear that particulate production from firewood and forest fires now amounts to a very small percentage of the anthropogenic particle production, which is dominated by the input from fossil fuels. In contrast, however, agricultural burning amounts to 20% of the anthropogenic input and windblown dust to a further 26%. Anthropogenic windblown dust is estimated to be larger than natural input of windblown dust; this could be exacerbated by poorly managed energy-plantation schemes. The prospect of converting biomass into liquid or gaseous energy carriers, efficient burning in stoves, etc., however, suggests that particulate emissions to the atmosphere due to biomass conversion schemes would not contribute significantly to presently estimated values of anthropogenic inputs to the atmosphere.

Table 6.4   The annual heat and water budget of Tunisian oases and the surrounding desert. Source: SMIC (1971)

	Oases	Semidesert
Area (km^2)	150	35,000
Estimated albedo	0.15	0.20
Bowen ratio	−0.26	5.6
Net radiation (W m^{-2})	100	80
Sensible heat transfer (W m^{-2})	136	12
Evapotranspiration (cm a^{-1})	168	15
Precipitation (cm a^{-1})	15	15

A further factor that may be taken into account with regard to biomass conversion is the possible need for irrigation in the case of energy-plantation schemes, especially since such schemes are likely to be relegated to marginal agricultural lands, since food production will require the better agricultural conditions. Munn and Machta (1979) point out that irrigation projects can be of local climatic significance, but the effects are sometimes difficult to detect on a larger scale. They suggest that, at a height of 10 m above a large agricultural area under irrigation or at a downwind distance of 1 km from a reservoir, the climatic effects may be imperceptible. However, it is possible for such changes in surface conditions to trigger cloud or shower development. Schickedanz (1976), cited by Munn and Machta (1979), has found higher rainfall in the vicinity of irrigated regions than in other regions during June, July, and August in Kansas, Nebraska, and most of Texas, whereas in April, May, and September when irrigation was not used, there was no anomaly. The effect of irrigation is, of course, most likely to be noticeable in arid regions. Table 6.4 shows the effect on the heat budget of the Sahara oases of southern Tunisia (SMIC, 1971), where a negative value of the Bowen ratio occurs, since the latent heat flux exceeds the available radiation. The increased atmospheric humidity in such desert regions is not likely to lead to increased precipitation because of the prevailing large-scale subsidence. This is confirmed by recent work using a numerical model of the mesoscale atmospheric circulation. It was used to investigate the mesoscale climatic changes due to a deliberate flooding of the Qattara depression in Egypt (Segal et al., 1982). Simulation of a typical summer synoptic situation indicated significant effects on the horizontal and vertical wind fields and on the temperature and moisture patterns. The large-scale subsidence due to the subtropical high would presumably limit any increase in precipitation.

(d) Wind-energy conversion

Wind power was once a more significant supplier of energy than it is today. Many new technologies are, however, being developed (e.g., vertical-axis windmills) and it is suggested that, with declining costs due to mass production and technological innovation, the use of wind power will increase rapidly, firstly in rural areas of the developing countries and then in the most wind-rich parts of the industrial world.

The IIASA Energy Systems Program (1981) estimated the technical potential of wind by considering the wind energy available at heights up to 200 m above ground level in continental regions within about 1,000 km of the coast between 50°N and 30°S. The Program's estimate of this technical potential is 3 TWyr/yr; however, because of constraints such as economics, aesthetics, and competing land uses, its estimate of the realizable potential of wind power is 1 TWyr/yr (see Table 6.1).

Reed (1980) calculates that the exploitable wind resource, in the atmospheric boundary layer is 20 TWyr/yr. For the United States, he estimates that an average of 7.5 TWyr/yr flows northwards in the boundary layer. He points out, however, that wind energy is not uniformly distributed in space or time, and obvious problems include operating wind-energy collectors over a wide dynamic range of inputs and smoothing outputs to conform to relatively uniform and regular demands. The potential for wind energy in certain areas is considerable. For example, Baker *et al.* (1979) have pointed out that the Pacific northwest area of the US is endowed with substantial resources of both hydro- and wind energy and that the combination of these into an integrated and optimized system has the potential for supplying a major portion of the future energy requirements of Oregon, Washington, Idaho, western Montana, and northeastern Nevada. An analysis of the wind-energy potential in the US (Elliot, 1978) shows that the annual wind power at 50 m above exposed areas is high (>400 W/m^2) over non-mountainous regions: the central and southern Great Plains, parts of Wyoming, Montana, and offshore and exposed coastal areas of the northeast, northwest, Alaska, and south Texas. Moderately high wind-power potential (>300 W/m^2) is indicated over the northern Great Plains, Great Lakes, and Hawaii. There are also isolated good sites throughout the rest of the country.

The main effect on local climate of wind-energy-conversion systems is likely to be on a microclimatological scale unless huge arrays of turbines are constructed, which seem unlikely. Hewson (1975), in assessing the effect of large-scale use of wind power, suggests that it is unlikely for any appreciable effect on climate to occur. The possible effects are illustrated by analogy with the growing of a number of groves of tall trees: in a wind the branches of a tree extract energy from the wind as do the rotating blades of a turbine. Hewson suggests that there might be a slight slowing of the wind for a short distance downwind from an array of windmills, but that the winds would rapidly accelerate because of downward transport of momentum from the stronger winds aloft.

(e) Hydropower

Hydropower is the most commonly used renewable source of electricity at the present. For example, in 1976 hydropower was the source of 73% of Canada's electricity. Weingart (1978) estimates that roughly 10% of the total global potential for hydropower has been tapped. Future development will, however, generally be resource limited in most developed countries and the greatest remaining potential is in Africa and Latin America. Both Weingart (1978) and Hayes (1978) point out that problems such as silting, flooding productive land, and reducing downstream fertility will probably limit the use of hydropower to about four times its present level. The IIASA Energy Systems Program (1981) estimates that 5 TWyr/yr of mechanical power are available from continental run-off of water, but for technical reasons only 3 TWyr/yr could be relied on. In view of reasons such as those listed above, however, IIASA estimates that the realizable potential of hydropower is 1.5 TWyr/yr (see Table 6.1).

Goldemberg (1980) has discussed the potential role of hydroelectric power in developing countries. He states that the world hydroelectric resources are very large (2.2 TWyr/yr of generating capacity at 50% capacity factor) and that most of this potential is in developing countries. These figures do not include waterfalls or dams of less than 20 m head or hydropotential smaller than 10 MW. Minihydro plants appear to have considerable potential and would, according to Goldemberg, increase the world hydroelectric potential to at least 3 TWyr/yr. Goldemberg uses figures from Brazil to emphasize the future role of hydroelectric power in developing countries. Only 14% of the Brazilian hydroelectric capacity is being used so far and hydroelectric production is increasing at an average rate of 11.3% per year in a government programme that will probably continue for the next 15 years. The power will be transmitted to consumption centres by high-voltage direct-current lines.

From the maximum realizable hydropower potential, no global-scale climate changes can be expected. On a local or regional scale changes in evaporation, cloud cover, and precipitation could occur. Such effects have also been mentioned in Section 6.2(c) with regard to the effects of large-scale irrigation.

(f) Wave, tidal, and ocean current energy

The IIASA Energy Systems Program (1981) points out that, from the total of 3 TWyr/yr of dissipated tidal power, a very small fraction is accessible for operating turbines; only a few coasts have a form that transforms the kinetic energy of the global tide into sufficiently large tidal levels. The Energy Systems Program estimates the technical potential of tidal power therefore to be about 0.04 TWyr/yr. The realizable potential would be even smaller, and applications would only be on a local scale. Thus, on the basis of the scale of use anticipated, no climatic effects can be expected from tidal energy conversion.

Wave power has been converted to mechanical power in some experimental systems. Weingart (1978) calculates that in order to produce 1 TWyr/yr, a wave machine operating in North Sea conditions and equal to the circumference of the earth would be required. Thus wave energy would appear to be an option which might be used in certain favoured locations but will not contribute significantly to a global energy supply of the order of tens of TWyr/yr.

The Energy Systems Program (1981) estimates that the total technical potential of ocean currents and waves is 0.005 TWyr/yr. It is pointed out that the kinetic energy of the ocean currents is 0.2 TWyr/yr, but, since the ocean currents play a significant role in the earth's heat balance by transporting energy poleward, it is felt that large-scale tapping of ocean-current energy should be avoided.

Von Arx, Stewart, and Apel (Stewart, 1974) estimate the effect on the Gulf Stream near the coast of Florida of extracting electrical power by using turbine arrays. They suggest that a reasonable turbine array could yield about 1,000 MW. However, as Baker (1977) has pointed out, the total kinetic energy of the current in this location is 25,000 MW and the turbine array would therefore be extracting at least 4% of the kinetic energy. The effect of such an extraction on Gulf Stream dynamics is not known, but could be important because of the potential sensitivity of the meandering path of such western ocean boundary currents to local changes and the possible importance of this path to air–sea exchange in subpolar regions. The poleward transport of energy by ocean currents plays an important role in the climate system; Vonder Haar and Oort (1973) estimate that in the region of maximum net northward energy transport (30°N to 35°N) the oceans transport 47% of the required energy. At 20°N the peak ocean transport accounts for 74% of the required energy transport at this latitude. Interference with the northward transport of energy by the oceans could have large-scale climatic effects.

It thus appears that wave and tidal energy-conversion systems will not be used on a large enough scale to have any climatic effects. Large-scale use of ocean-current energy could have a significant impact on global climate, but, since the technical potential is estimated to be only 0.2 TWyr/yr, large-scale use would seem to be unlikely.

(g) Ocean thermal electric conversion

Ocean thermal electric conversion (OTEC) plants use the vertical temperature difference in ocean water to drive turbines and generate electricity. Since the temperature difference is generally small (between ∼20 °C at the ocean surface and ∼5 °C in deep water), the efficiency of OTEC systems is low (2–3%). The Energy Systems Program (1981) calculates the technical potential of OTEC to be 22 TWyr/yr of electricity and points out that this would involve the transfer of 720 TWyr/yr from the ocean surface layers to deeper layers. If this heat were subtracted from that which is usually transported north, then, as the Energy Systems Program suggests, the climatic effects would be significant. On the other hand, the technical potential of OTEC systems is estimated to be only 1 TWyr/yr if the potential of the areas of natural transfers of cold water from depths to the surface (i.e., areas of ocean upwelling) are considered. Since OTEC has extremely speculative economics and is in a very early stage of development, the realizable potential is estimated by IIASA to be 0.5 TWyr/yr of electricity.

Weingart (1978) claims that the theoretical potential of OTEC systems is more than 100 TW(th) but points out that most of this potential is at deep ocean sites too distant for electricity transmission to shore. If the latter were the only constraint, one could envisage OTEC systems being used to produce a transportable secondary energy carrier such as hydrogen.

Lavi and Lavi (1980) have discussed the development of OTEC systems and suggest that the global market potential may exceed 1 TW, but that constraints will reduce this. It is suggested that, in the US, OTEC plants in the 10–100 MW range would be economical for US islands depending on imported oil and that OTEC could enter the mainland market in the southeast if projected capital costs for large plants are realized and if high-voltage underwater DC transmission is developed. Lavi and Lavi (1980) point out, however, that, even if it is assumed that OTEC is technically and economically sound, there are many factors that could constrain large-scale penetration into the US mainland market, such as capital investment requirements and institutional and environmental questions.

Zener (1973) calculated that by siting OTEC plants all through the oceans between 20°N and 20°S a total of 60 TWyr/yr could be generated and estimated that this would result in a persistent 1 °C decrease in the ocean surface temperature over this zone. Harrenstein and McCluney (1976) calculate that OTEC plants would lower the surface temperature of the Gulf Stream by 0.5 °C if there were enough plants to generate the current US electrical demand of $\sim 2 \times 10^5$ MW.

The interactions between the atmosphere and the ocean are of major importance within the climate system. Observations and model studies have indicated that ocean surface temperature anomalies can have a significant effect on the overlying atmospheric circulation and sometimes on the atmospheric circulation in other regions. Such studies have been discussed in Chapter 2 (Section 2.8). This climate sensitivity to ocean temperature anomalies must be borne in mind when large-scale use of OTEC is considered.

In addition to the effects on the ocean temperature distribution and kinetic energy of ocean currents, the influence of OTEC systems on climate could arise through secondary effects from the transfer of colder deep ocean water to the surface layers. Firstly, the colder deep ocean water is enriched in CO_2 and von Hippel and Williams (1975) suggest that the release of this water at the surface would imply that an OTEC plant would release about one-third as much CO_2 to the atmosphere as does a fossil-fuel power plant with an equivalent energy production. Secondly, the deeper ocean water has an enhanced nutrient supply and it is therefore possible that the release of this water near the surface would lead to phytoplankton blooms and thus possibly to widespread ocean-surface albedo changes. Such albedo changes, related to phytoplankton blooms, have been discussed in another context by Siemerling (see, for instance, AIHS, 1978), who has suggested that an increased atmospheric CO_2 level would fertilize the surface ocean and the resulting albedo changes would cause a global cooling.

(h) Solar satellite power systems

The concept of solar power satellites (SPS) is that they use photovoltaic cell arrays to intercept solar radiation in space; the resulting electricity is converted to microwaves, which are beamed to the earth's surface. The advantage of having collectors in space is that a given area intercepts four times as much radiation there as in the sunniest spot on earth, since there is no overlying atmosphere, no weather and no day–night cycle. Ground equipment reconverts the microwave energy to electricity. An SPS system has been envisaged (Lewis, 1979) with about 100 satellites in geostationary orbit. A 5 GW SPS now being studied has a rectangular array of photovoltaic cells, 10.5 km long and 5 km wide. The receiving antenna at the earth's surface would be 10 km × 15 km in area (Moses, 1980). A system of 100 SPS would also require 10 launch sites, each covering an area of about 20 km². The major cost of SPS systems would be the rocket launches. An SPS would be heavy (18,000 t according to Ridpath, 1978) and supershuttles, capable of lifting 500 t into orbit at a time would be required (the present shuttle lifts some 29 t). A continuous run of rocket launches, perhaps one a day for years on end would have to be considered. Such technological and economical constraints together with certain environmental constraints (effects of microwave beam, impact of space shuttle exhausts on ozone layer) will probably limit this option.

Moses (1980) has evaluated the effect of SPS on the atmosphere. He lists five types of atmospheric problems: (i) the dispersion of effluents from thruster motors; (ii) noise; (iii) microwave phenomena in the atmosphere (e.g., heating or ion depletion); (iv) fugitive emissions such as waste heat, dust, or debris; and (v) the large metal structures in orbit and on the ground.

In the upper atmosphere, the effluent from vehicles may cause a 50% reduction in the ionization of the F layer and air glow. Other effects include the increase in condensation nuclei with subsequent effects on the troposphere and changes in the electric field. Waste heat generation, dust clouds, and changes in the environment of particles could also have climatic effects (Moses, 1980).

In the troposphere, the microwave effects would result in the heating of rain, snow, and ice particles and this effect could influence cloud dynamics as well as the atmospheric electric fields and frequency of lightning. Moses (1980) estimates that the waste heat associated with the receiving antenna would have roughly the same order of magnitude as that of an urban heat island, the effects of which have been discussed in Chapter 4.

Carbon dioxide, water vapour, and oxides of nitrogen would be injected into the stratosphere and upper atmosphere. The effect of water vapour at 80 km was estimated by Moses (1980) to be quite serious, whereas the effects of CO_2 or nitrogen oxides were estimated to be small.

6.3 POSSIBLE EFFECTS ON A GLOBAL SCALE

Emphasis in the last section was on climatic effects on the local or regional scale, although some potential global effects with regard to OTEC and solar satellite power systems were mentioned. In general, as indicated in the last section, the renewable energy systems are most likely to be applied in small- to medium-sized, decentralized or widely distributed units. Thus, it seems unlikely that they will cause perturbations to the climatic boundary conditions large enough to force changes in the atmospheric circulation. As shown, however, in Chapter 4, global-scale effects are possible if the man-made perturbations are large enough and particularly if they are in sensitive areas of the climate system.

Since renewable energy-conversion technologies are generally in an early stage of development, still mostly in the pilot-scheme stage, or even in the research stage, it is clear that deployment on a scale large enough potentially to influence global climate will in any case occur in the distant future. Moreover, the detailed physical characteristics of the systems are not well known. Thus, it is unrealistic to examine specific scenarios for the large-scale use of these technologies. The approach in this section is therefore to describe the general types of effect that could arise, often without reference to specific systems. For example, STEC, photovoltaic, and biomass systems could each cause large-scale changes in the surface energy balance. Therefore model and observational studies that investigate the effect of large-scale albedo changes are discussed in order to illustrate the types of feedback that are involved rather than the effects of a particular system. The one exception to this is the description of a model study of the global effect of STEC systems.

(a) The effect of large-scale changes in the surface energy balance

Most studies of the effect of large-scale changes in the surface energy balance have considered the influence of changes in the surface albedo, that is, the effect of large-scale changes in the amount of solar radiation absorbed at the earth's surface. On a global scale, Hummel and Reck (1978), as cited by Munn and Machta (1979), conclude that, if the amount of arable land were increased by 1% and its albedo changed from that of black soil (0.07) to that of crops (0.25) for one-third of the year, the global average surface temperature would be decreased by 1 K. Likewise, man-made lakes and reservoirs are estimated to have increased the surface temperature by 0.4 °C. Munn and Machta (1979) point out that albedo changes are concentrated in only a few regions of the earth, so that the climatic effects would in the first instance be on a local or regional scale. Flohn (1975) calculated, using the one-dimensional radiative–convective model of Manabe and Wetherald (1967), the equilibrium temperature deviations for different values of average albedo corresponding to periods in the past. Table 6.5, derived from Flohn (1975), shows the computed deviations. In terms of the global

Table 6.5 The equilibrium temperature deviations due to albedo changes, calculated by the model of Manabe and Wetherald (1967). Source: Flohn (1975)

Present albedo (1901–1950)	N. Hemisphere 0.1294 S. Hemisphere 0.1384	
	Albedo	Temperature deviation from present (K)
N. Hemisphere 1890	0.1373	−0.95
S. Hemisphere 1850	0.1434	−0.60
Ice Age	0.1731	−4.6
Ice Age (sea level −100 m)	0.1860	−6.2
Antarctic ice surge	0.2022	−7.6

average surface temperature, we see that large-scale albedo changes could have a significant effect. Thus the large-scale use of photovoltaic power plants or biomass plantations, if they produced appropriate albedo changes, could have a similar influence.

The effect of large-scale changes in albedo on the atmospheric circulation has been studied in a number of model experiments, which began when scientists were attempting to explain the origins of the Sahel drought. Charney (1975) proposed a biogeophysical feedback mechanism to explain the drought. It was argued that a reduction of the vegetation cover (initially because of overgrazing) would increase the albedo in the Sahel region and this would lead to an increase in atmospheric sinking motion giving an additional drying, which would perpetuate the arid conditions. To test this hypothesis, Charney et al. (1975) used the GCM developed at the Goddard Institute for Space Studies to compare the atmospheric circulation simulated when the albedo in the Sahel area was 0.14 and 0.35. A decrease in precipitation and convective cloud cover was reported as a response to the increased albedo. A further series of experiments by Charney et al. (1977) took better account of the interaction between surface hydrological processes, albedo changes, and the atmospheric circulation. It was found that, in an area where appreciable evaporation from the surface occurs, an increase of albedo reduces the absorption of solar radiation by the ground and consequently the flux of sensible and latent heat into the atmosphere. The resulting reduction in convective cloud cover then tends to compensate for the increase in albedo by allowing more solar radiation to reach the ground, but it simultaneously reduces the downward flux of long-wave radiation even more, so that the net absorption of radiation by the ground is decreased. Without evaporation, the increase of albedo causes a decreased absorption of solar radiation and thus a decrease of convective cloud cover and precipitation.

A further test of Charney's idea was made by Ellsaesser et al. (1976) using a two-dimensional zonal atmospheric model. In this case, two model runs were made: one with an albedo of 35% over the Sahara and one with an albedo of 14%. For the zonal model, these albedos were inserted over the fraction (30%) of the latitude band encompassed by the Sahara (15°–25°N in the model). The results showed that the increase in albedo caused a reduction of precipitation of 22% over land at 20°N.

Chervin (1979) has made similar experiments with a version of the NCAR GCM. Figure 6.8 shows the locations over which the albedo was increased to 45%. In the control cases, the albedo varied geographically between 7% and 17% for the US high-plains area and between 8% and 35% for the Sahara region. Figure 6.9 shows the changes in precipitation rate produced in the two simulations. We see that large and statistically significant decreases in precipitation are found over the prescribed change region together with areas of significant increases to the north and south. For two rows of grid points south of the altered region in Africa the precipitation is significantly decreased. In general, Chervin finds that the model response is a local one, except for a physically self-consistent and statistically significant change over the equatorial Indian Ocean just west of

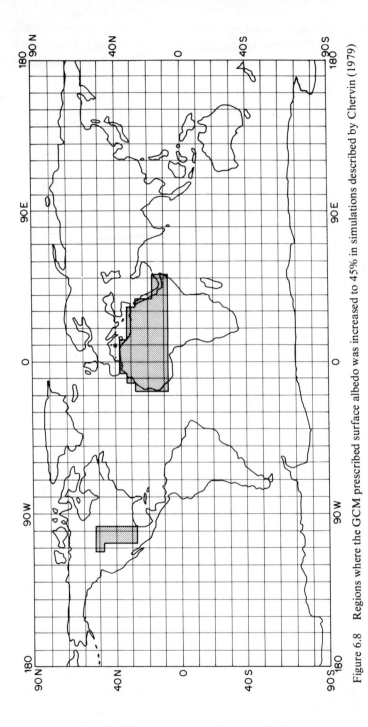

Figure 6.8   Regions where the GCM prescribed surface albedo was increased to 45% in simulations described by Chervin (1979)

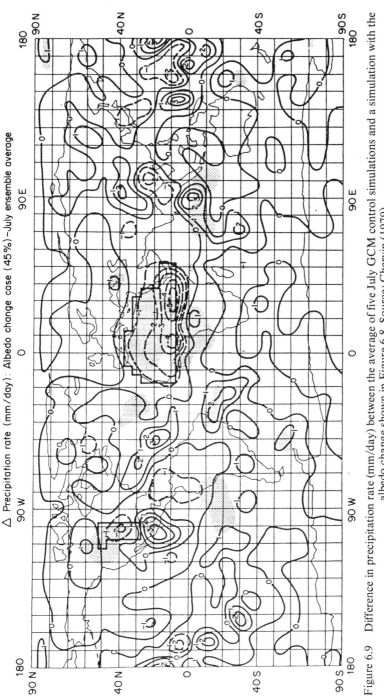

△ Precipitation rate (mm/day): Albedo change case (45%) –July ensemble average

Figure 6.9 Difference in precipitation rate (mm/day) between the average of five July GCM control simulations and a simulation with the albedo change shown in Figure 6.8. Source: Chervin (1979)

Sumatra, where there is a precipitation increase as shown in Figure 6.9. The results suggest that for a large enough albedo increase the surface evaporation can be reduced enough to cause a surface temperature increase. In terms of a global average response, it was found that a significant cooling occurred throughout the model atmosphere; however, the globally averaged precipitation response was not significantly different, in spite of the large significant regional changes. Apparently the regional changes compensated each other to such an extent that there was essentially no change in the global mean precipitation.

Potter *et al.* (1975) describe three integrations made with a zonal atmospheric model to investigate the possible climatic effect of tropical deforestation. In the control run the rainforest albedo was 0.07; the complete removal of the tropical rainforest was simulated by increasing the albedo to 0.25, combined with an increased run-off rate and decreased evaporation rate; 'wet' deforestation was simulated by increasing the albedo to 0.25 only. Figure 6.10 shows the change of precipitation in the area of albedo increase and an increase of precipitation to the north and south. The chain of consequences following the deforestation was suggested to be: deforestation → increased surface albedo, reduced surface absorption of solar energy → surface cooling → reduced evaporation and sensible heat flux from the surface → reduced convective activity and rainfall → reduced release of latent heat, weakened Hadley circulation, cooling in middle and upper tropical troposphere → increased tropical lapse rates → increased precipitation in the latitude bands 5°–25°N and 5°–25°S and decreased equator–pole temperature gradient, reduced meridional transport of heat and moisture out of equatorial regions → global cooling and decrease in precipitation between 45° and 85°N and 40° and 60°S.

A further experiment with the same model has been reported by Potter *et al.*

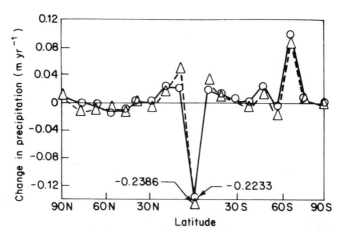

Figure 6.10   Difference in precipitation (perturbed minus control simulations) for case simulating complete tropical deforestation (circles) and case with albedo increase in tropical areas (triangles).
Source: Potter *et al.* (1975)

(1980). In this case, the effects of a combination of desertification of the Sahara and deforestation of the tropical rainforest were studied. The most significant model responses were found in the modified zones, although the high northern latitudes exhibited the greatest cooling as a result of the ice–albedo feedback. Potter *et al.* concluded that anthropogenic modification of the surface albedo during the past few thousand years has had an effect on global climate that is probably quite small and undetectable.

One preliminary analysis has been made by Potter and MacCracken (1977) to evaluate the effect on climate of the large-scale use of STEC systems using the two-dimensional (latitude and height) zonal atmospheric model. A scenario for albedo modification due to large-scale use of STEC systems, derived by Grether *et al.* (1977), was used as an input to the model. This scenario assumed a world population of 10,000 million with a *per capita* energy requirement of 10 kW. It was further assumed that to generate 100 MW(e) a reflector area of 3 km^2 and a total land area of 9 km^2 would be required. It was suggested that the albedo change due to STEC plants would be such that an area equal to the total reflector area would be completely black. Thus the new albedo was assumed to be two-thirds of the natural albedo where the STEC plants were located. Figure 6.11 shows the land area that Grether *et al.* (1977) assumed to be devoted to STEC facilities.

Using this scenario, Potter and MacCracken (1977) modified the zonally averaged albedo in the model within the zone 50°N to 40°S. No other boundary conditions were altered (e.g., run-off, evaporation, and surface roughness), even though these could also be influenced by large-scale use of STEC. Figure 6.12 shows the latitude–height distribution of change in atmospheric temperature

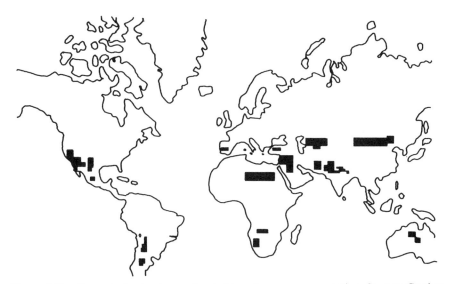

Figure 6.11   Scenario for land area devoted to solar energy conversion. Source: Grether *et al.* (1977)

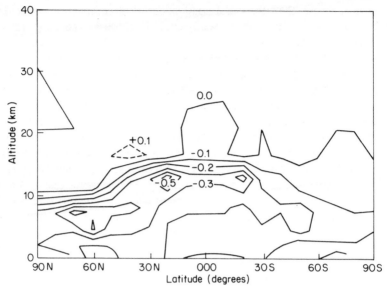

Figure 6.12 Latitude–height distribution of difference in temperature between model case considering large-scale solar energy conversion and control case (°C). Source: Potter and MacCracken (1977)

resulting from the albedo changes. The troposphere warmed because of increased absorption of solar radiation, with the maximum warming at the latitudes of largest coverage of STEC plants. Precipitation increased in the tropics and subtropics.

The results of the study made by Potter and MacCracken should be considered as preliminary for two reasons. Firstly, the scenario investigated has several limitations, not only because of the large amount of solar energy conversion that it assumes, but also because of the simplifying assumptions made regarding the effective changes in boundary conditions. As outlined in Section 6.2(a), STEC facilities are more likely to cause a redistribution of the surface energy-balance components rather than a change of albedo. Secondly, the effects of a zonally averaged albedo change within a zonally averaged model were investigated and it is difficult to extrapolate these results in order to discuss the effects of regional energy-balance changes on the general atmospheric circulation. Nevertheless, the results show that large-scale albedo changes can influence the atmospheric temperature, precipitation distribution, and other climate variables, not only over the zones of albedo change, but elsewhere through a number of atmospheric feedback mechanisms.

(b) The effect of large-scale changes in surface roughness and hydrological characteristics

The surface roughness influences the surface drag, energy dissipation, and turbulent exchange of the air. Baumgartner (1979) indicates that world maps and simulations of coverage show a considerable influence of forest cover on the above variables and on the angle of deviation of the geostrophic wind from isobars and through such interactions, large-scale changes in forest cover could influence the development of the pressure field and the circulation of the atmosphere. Thus, a potential effect on global climate could occur through the changes in surface roughness brought about by large-scale use of biomass energy conversion (introduction of plantations), STEC (heliostats several metres tall), or photovoltaic energy conversion (flat-plate or concentrating collectors).

Few model studies have been made to investigate the global climatic effect of surface roughness changes and none made with specific reference to energy-conversion systems (Williams, 1977). In atmospheric GCMs the drag coefficient C_D is used in computations of the horizontal stress components and the vertical fluxes of sensible heat and water vapour in the surface boundary layer. Two cases have been run with the NCAR GCM in which C_D was changed. When C_D was doubled, the surface winds were reduced but the upper winds were increased, because there was an increased surface moisture flux in the tropics, giving an increased poleward momentum transport and stronger upper westerlies. Further indications of the interactions involved are shown in a series of GCM cases run by Delsol et al. (1971), who pointed out the complexity of the interactions involving changes in C_D. For example, it was suggested that an increase in C_D contributes to filling depressions and therefore to reduced precipitation. On the other hand, an increase in C_D intensifies upward currents, including the upward flux of water vapour at the top of the surface boundary layer, and this could favour an increase in precipitation.

Large-scale changes in surface hydrological characteristics could have a significant effect on climate. Results of GCM studies by Charney et al. (1977), described above, have already emphasized this point. Observations of the synoptic-scale effect of large-scale changes in surface wetness are few. However, Namias (1960) has discussed the influence of abnormal heat sources and sinks on atmospheric behaviour and in particular has discussed reasons for the persistence of drought in the Great Plains area of the US. The occurrence of an upper-level anticyclone with dry subsiding air is characteristic for drought in this area. Between spring and summer there is a tendency for such an anticyclone to develop over the Southern Plains, but there are years when the anticyclone does not become established. Studies of observed data showed that in years when the anticyclone was strong and persistent in the summer, there was a positive anomaly in the contour patterns in the spring. Namias has suggested that soil moisture conditions play an important role in this correlation. Thus, when the Southern Plains have been dominated in the spring by a very dry regime, which is usually very warm, and the soil is desiccated, then the opportunity for persistent lodgement of the upper level anticyclones in the summer is favoured. On the other hand, Namias proposed that after a wet spring some of the heat normally used to raise the temperature of the ground surface might be used to evaporate the excess water in and on the soil and thus not be available for sensible heating of the air, which could be necessary for the maintenance of the upper level anticyclone.

Thus it appears possible that large-scale anomalies in surface wetness could influence the overlying atmospheric circulation through changes in heating rates. Persistent anomalies in the large-scale atmospheric circulation could lead to effects on the upstream or downstream flow, giving anomalies in other areas.

Walker and Rowntree (1977) have examined the sensitivity of a model of the tropical atmospheric circulation to changes in soil moisture content. The model area considered a zonally symmetric version of West Africa. In one simulation there was a desert in the same latitudes as the Sahara; in the second simulation the desert was replaced by wet land. Results suggested that once the land was moist it maintained itself in this state for at least several weeks, whereas initial aridity north of 14°N was sustained, suggesting that ground dryness alone can cause deserts to persist. This has implications for any scheme which would involve large-scale drying out of the surface (e.g., paving for STEC or photovoltaic installations), or large-scale wetting of the surface (e.g., irrigation for plantations).

Barry and Williams (1975) noted that the absolute values of precipitation computed by the NCAR GCM were, particularly in the areas that actually are deserts, almost consistently too large, primarily because of the assumption, in that version of the model, of a Bowen ratio (see Section 5.2.4) equal to unity. In a simulation using a Bowen ratio equal to 10, which corresponds to greatly reduced evaporation, the values of released latent heat (and presumably, therefore, of precipitation) were more realistic over the arid areas of the world. This again illustrates the large-scale effects of surface hydrological characteristics.

6.4   CONCLUDING REMARKS

Since many of the solar energy conversion schemes are likely to be deployed only on a small scale in isolated regions, in which case any very localized effect would be more dependent on particular local conditions, the local effect of only four systems need be considered at the present time. Solar thermal-electric conversion power plants could influence local climate through changes in the surface energy balance, in the surface roughness, and in the surface hydrological characteristics. In reality, such changes would be complex. Rough estimates suggest, however, that the effects on the surface energy balance would not lead to a change in the amount of heat added to the atmosphere but in the manner in which it is added. The effect of such energy-balance changes has been investigated by using a two-dimensional mesoscale model of the atmosphere (Bhumralkar et al., 1979). It was found that a hypothetical STEC plant in southern Spain, generating about 30 GW, increased local cloudiness and precipitation in summer simulations.

Photovoltaic cells could be deployed in a number of ways ranging from arrays of cells on the roofs of houses to modules of cell arrays for a central power station. At present, the use of photovoltaic conversion in a decentralized fashion appears to be the preferred option, in which case local/regional effects would not differ in magnitude from those of urban areas. An estimate of the effect of larger-scale flat-plate collector systems suggests that, if they were installed in an area where the surface was naturally dark (e.g., forest, wet fields), then there would be no significant change in the amount of energy input to the atmosphere. If cells with extremely low efficiency were used and the natural area had a high reflectivity, then the photovoltaic system would act as a heat source. However, reference systems presently under discussion (e.g., EPRI, 1978) would have areas on the order of only 3 km² and for this size the percentage change in the amount of energy input would probably be negligible.

Biomass can be converted to solid, liquid, and gaseous fuels from numerous sources and via many technologies. This heterogeneity means that no simple evaluation of the local and regional effects of biomass conversion can be given. The study for a 'Solar Sweden' proposed that half of the energy required in the year 2015 in Sweden would be from biomass, requiring energy plantations of 3 million hectares (Johansson and Steen, 1978). However, the forestry industry in Sweden presently uses 23 million hectares; thus the effect of such energy conversion would be smaller than that of present forests. Likewise, biomass conversion schemes to use firewood in the future are based on technology to use the wood more efficiently than at present, thus reducing the effect. Forests influence the exchange of water between the earth's surface and the atmosphere, consuming more water than other vegetation and transferring it by evapotransportation back into the atmosphere (Baumgartner, 1979). Forests also have a dynamic interaction with the atmosphere, since the forest is a rough earth–atmosphere interface, which influences heat and momentum transfer. Particle emissions to the atmosphere due to biomass conversion schemes would probably not contribute significantly to the presently estimated values of anthropogenic inputs to the atmosphere.

The main effect on local climate of wind-energy conversion systems is likely to be on a microclimatological scale unless huge turbine arrays are considered, which seems unlikely. Hewson (1975) suggests that there might be a slight slowing of the wind for a short distance downwind from an array of windmills, but that the wind would rapidly accelerate because of downward transport of momentum from the stronger winds aloft.

Since most renewable energy-conversion schemes that could be used on a large scale are still in the development stage, it is clear that their deployment on a scale large enough to influence potentially the globally averaged climate will, in any case, occur in the distant future, if at all. Moreover, the detailed physical characteristics of potential large-scale systems are not well defined. It is therefore unrealistic to examine specific scenarios for using these technologies. It is possible, however, to describe the general types of effect that could arise from large-scale changes in surface energy balance, roughness, and hydrological characteristics (resulting, for instance, from large-scale use of STEC, photovoltaic, or biomass plantation systems) or large-scale changes in ocean temperature due to OTEC systems. A number of model studies have indicated that unrealistically large-scale increases in surface albedo can lead to increased cloudiness and rainfall over the perturbed area with further effects on the atmospheric circulation. Large-scale roughness changes could influence the atmospheric circulation owing to changes in the transfers of momentum, heat, and moisture in the boundary layer. Large-scale anomalies in surface wetness could lead to effects on the upstream and downstream flow, giving anomalies in other areas. Model and observational studies indicate that large-scale ocean-surface-temperature anomalies could have a significant influence on the atmospheric circulation.

REFERENCES

AIHS (1978). The consequences of a hypothetical world climate scenario based on an assumed global warming due to increased carbon dioxide. *Proc. of Symposium. Aspen Institute for Humanistic Studies*. Aspen, Colorado.

Arndt, W., G. Bilger, W. H. Bloss, G. H. Hewig, F. Pfisterer, and H. W. Schock (1978). CdS–Cu$_2$S thin film solar cells for terrestrial applications. In, *Photovoltaic Solar Energy Conference*. Reidel, Dordrecht, Holland.

Bach, W. (1976). Global air pollution and climatic change. *Rev. Geophys. Space Phys.*, **14**, 429–474.

Bach, W., J. Pankrath, and J. Williams (eds.) (1980). *Interactions of Energy and Climate*. D. Reidel Publ. Co., Dordrecht, Holland.

Baker, D. J. (1977). Ocean dynamics and energy transfer: Some examples of climatic effects. In, *Energy and Climate*. National Academy of Sciences, Washington, DC.

Baker, R. W., E. W. Hewson, N. G. Butler, and E. J. Warchol (1979). Wind power potential in the Pacific Northwest. *J. Appl. Meteor.*, **17**, 1814–1826.

Barry, R. G., and R. J. Chorley (1968). *Atmosphere, Weather and Climate*. Methuen, London.

Barry, R. G., and J. Williams (1975). Experiments with the NCAR global circulation model using glacial maximum boundary conditions: Southern hemisphere results and interhemispheric comparison. In, R. P. Suggate and M. M. Cresswell (eds.), *Quaternary Studies*. The Royal Society of New Zealand, Wellington.

Baumgartner, A. (1979). Climate variability and forestry. Proceedings of the World Climate Conference, WMO Publication No. 537, World Meteorological Organization, Geneva.

Berkofsky, L. (1976). The effect of variable surface albedo on the atmospheric circulation in desert regions. *J. Appl. Meteor.*, **15**, 1139–1144.

Bhumralkar, C. M., J. Williams, and A. Slemmons (1979). The impact of a conceptual solar thermal electric conversion plant on regional meteorological conditions: A numerical study. *Solar Energy*, **23**, 393–403.

Black, J. F. (1963). Weather control: Use of asphalt coating to tap solar energy. *Science*, **139**, 225–227.

Charney, J. G. (1975). Dynamics of desert and drought in the Sahel. *Quart. J. Roy. Meteor. Soc.*, **101**, 193–202.

Charney, J. G., P. H. Stone, and W. J. Quirk (1975). Drought in the Sahara: A biogeophysical feedback mechanism. *Science*, **187**, 434–435.

Charney, J. G., W. J. Quirk, S.-H. Chow, and J. Kornfield (1977). A comparative study of the effects of albedo change on drought in semiarid regions. *J. Atmos. Sci.*, **34**, 1366–1385.

Chervin, R. M. (1979). Response of the NCAR general circulation model to changed land surface albedo. *Report of the JOC Study Conference on Climate Models: Performance, Intercomparison and Sensitivity Studies*. GARP Publications Series No. 22, World Meteorological Organization, Geneva.

Delsol, F., K. Miyakoda, and R. H. Clarke (1971). Parameterized processes in the surface boundary layer of an atmospheric circulation model. *Quart. J. Roy. Meteor. Soc.*, **97**, 181–208.

Dittberner, G. J. (1978). Climatic change: volcanoes, man-made pollution, and carbon dioxide. *IEEE Trans. on Geoscience Electronics*, GE-16, 50–61.

Elliott, D. L. (1978). An overview of the national wind energy potential. *Proceedings of*

Conference on Climate and Energy: Climatological Aspects and Industrial Operations. American Meteorological Society, Boston, Massachusetts.

Ellsaesser, H. W., M. C. MacCracken, G. L. Potter, and F. M. Luther (1976). An additional model test of positive feedback from high desert albedo. *Quart. J. Roy. Meteor. Soc.,* **102**, 655–666.

Energy Systems Program (1981). *Energy in a Finite World: Volume 1. Paths to a Sustainable Future; Volume 2. A Global Systems Analysis.* Ballinger Publishing Company, Cambridge, Massachusetts.

EPRI (1976). Closed cycle high-temperature central receiver concept for solar electric power. EPRI ER-403-SY, Electric Power Research Institute, Palo Alto, California.

EPRI (1978). Perspectives on utility central station photovoltaic applications. EPRI ER-589-SY, Electric Power Research Institute, Palo Alto, California.

Flohn, H. (1975). History and intransitivity of climate. In, *The Physical Basis of Climate and Climate Modelling.* GARP Publications Series No. 16, World Meteorological Organization, Geneva.

Goldemberg, J. (1980). Global options for short-range alternative energy strategies. In, W. Bach *et al.* (eds.), *Renewable Energy Prospects.* Pergamon Press, Oxford, England.

Grether, D., M. Davidson, and J. Weingart (1977). A scenario for albedo modification due to intensive solar energy production. In, J. Williams, G. Krömer, and J. Weingart (eds.), *Climate and Solar Energy Conversion.* CP-77-9, International Institute for Applied Systems Analysis, Laxenburg, Austria.

Hall, D. O., 1979: Solar energy use through biology—past, present and future. *Solar Energy,* **22**, 307–328.

Harrenstein, H. P., and W. R. McCluney (1976). Gulf Stream OTEC resource potential and environmental impact assessment overview. In, *Proc. Joint Conference: American Section of International Solar Energy Society and Solar Energy Society of Canada.* Winnipeg, Canada.

Hayes, D. (1977). Plant power: Biological sources of energy. *The Ecologist,* **7**, 340–356.

Hayes, D. (1978). The solar energy timetable. Worldwatch paper 19, Worldwatch Institute, Washington, DC.

Hewson, E. W. (1975). Generation of power from the wind. *Bull. Amer. Meteor. Soc.,* **56**, 660–675.

Hildebrandt, A. F., and L. L. Vant-Hull (1977). Power with heliostats. *Science,* **197**, 1139–1146.

Hummel, J. A., and R. A. Reck (1978). Development of a global surface albedo model. GMR-2607, General Motors Research Labs, Michigan.

Jäger, F., S. Chebotarev, and J. Williams (1978). Large-scale deployment of solar thermal electricity generation in European countries. Systems aspects concerning market penetration, reliability and climate. In, *Solar Thermal Power Stations.* DFVLR, Cologne.

Johansson, T. B., and P. Steen (1978). *Solar Sweden.* Secretariat for Future Studies, Sweden.

Kelly, H. (1978). Photovoltaic power systems: A tour through the alternatives. *Science,* **199**, 634–643.

Lavi, A., and G. H. Lavi (1980). Ocean thermal energy conversion (OTEC) social and environmental issues. In, W. Bach *et al.* (eds.), *Renewable Energy Prospects.* Pergamon Press, Oxford, England.

Lewis, R. W. (1979). Solar satellites fly on nuclear misfortunes, *New Scientist,* May, 710.

Manabe, S., and R. T. Wetherald (1967). Thermal equilibrium of the atmosphere with a given distribution of relative humidity. *J. Atmos. Sci.,* **24**, 241–259.

Martin Marietta (1975). Central receiver solar thermal power system—Phase I. Preliminary design report. Dept. of Energy, Washington, DC.

Moltschanov, A. A. (1966). Peculiarities of the hydrology of catchment basins and the determination of the optimum land use. *Proceedings of the World Forest Cong.,* Madrid.

Moses, H. (1980). Impacts of satellite power system technology. In, W. Bach *et al.* (eds.), *Renewable Energy Prospects*. Pergamon Press, Oxford, England.

Munn, R. E., and L. Machta (1979). Human activities that affect climate. *Proc. of the World Climate Conference*. WMO Publication No. 537, World Meteorological Organization, Geneva.

Namias, J. (1960). Influences of abnormal heat sources and sinks on atmospheric behavior. *Proc. Intern. Symp. Numerical Weather Prediction, Tokyo*. Meteor. Soc. Japan, pp. 516–629.

Potter, G. L., H. W. Ellsaesser, M. C. MacCracken, and F. M. Luther (1975). Possible climatic impact of tropical deforestation. *Nature*, **258**, 697–698.

Potter, G. L., and M. C. MacCracken (1977). Possible climatic impact of large-scale solar thermal energy production. In, J. Williams, G. Krömer, and J. Weingart (eds.), *Climate and Solar Energy Conversion*. CP-77-9, International Institute for Applied Systems Analysis, Laxenburg, Austria.

Potter, G. L., H. W. Ellsaesser, M. C. MacCracken, J. S. Ellis, and F. M. Luther (1980). Climate change due to anthropogenic surface albedo modification. In, W. Bach, J. Pankrath, and J. Williams (eds.), *Interactions of Energy and Climate*. Reidel, Dordrecht, Holland.

Reed, J. W. (1980). An analysis of the potential of wind energy conversion systems. In, W. Bach *et al.* (eds.), *Renewable Energy Prospects*. Pergamon Press, Oxford, England.

Ridpath, I. (1978). Sunny future for power satellites. *New Scientist*, 25 May, 520–522.

Robinson, G. D. (1977). Effluents of energy production: Particulates. In, *Energy and Climate*, National Academy of Sciences, Washington, DC.

Schickedanz, P. T. (1976). The effect of irrigation on precipitation in the Great Plains. NSF Gl-43871, Illinois State Water Survey, University of Illinois, Urbana, Illinois.

Segal, M., R. A. Pielke, and Y. Mahrer (1982). On climatic changes due to a deliberate flooding of the Qattara depression (Egypt). *Climatic Change*, **5**, 73–84.

SMIC (1971). *Inadvertent Climate Modification*. Report of the study of man's impact on climate. MIT Press, Cambridge, Massachusetts.

Stewart, R. B. (ed) (1974). *Proceedings of the MacArthur Workshop on the Feasibility of Extracting Usable Energy from the Florida Current*. Palm Beach Shores, Florida.

Vonder Haar, T. H., and A. Oort (1973). New estimate of annual poleward energy transport by northern hemisphere oceans. *J. Phys. Oceanogr.*, **3**, 169–172.

von Hippel, F., and R. H. Williams (1975). Solar technologies. *Bull. Atomic Scientist*, **31**, 25–30.

Walker, J., and P. R. Rowntree (1977). The effect of soil moisture on circulation and rainfall in a tropical model. *Quart. J. Roy. Meteor. Sci.*, **103**, 29–46.

Weingart, J. M. (1978). Solar energy as a global energy option for mankind. In, *Solar Thermal Power Generation*. UK-ISES, London.

Williams, J. (1977). Experiments to study the effects on simulated climate of changes in albedo, surface roughness, and surface hydrology. In, J. Williams, G. Krömer, and J. Weingart (eds.), *Climate and Solar Energy Conversion*. CP-77-9, International Institute for Applied Systems Analysis, Laxenburg, Austria.

Williams, J., and Krömer, G. (1979). A systems study of energy and climate. SR-79-2B. International Institute for Applied Systems Analysis, Laxenburg, Austria.

Williams, J., G. Krömer, and J. Weingart (eds.) (1977). *Climate and Solar Energy Conversion*. CP-77-9, International Institute for Applied Systems Analysis, Laxenburg, Austria.

Zener, C. (1973). Solar sea power. *Physics Today*, 48–53.

CHAPTER 7

The Effect of Climate on Energy Supply and Demand

7.1 INTRODUCTION

Chapters 2–6 have discussed the effect of energy conversion on the climate system. The reverse interaction, namely the effect of climatic variations on energy supply and demand, has not received as much attention in the scientific literature. However, recent increasing awareness of society's dependence on climate and vulnerability to climatic change suggests that the topic will be more seriously considered in the future (e.g., WCC, 1979).

Figure 7.1, from Critchfield (1978), shows the relations between climate and energy policy considerations. Obviously climate has an effect on demand for energy; studies referred to later in this chapter have shown the effect of cold

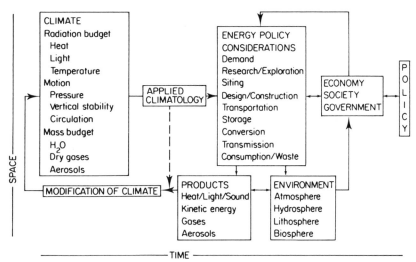

Figure 7.1   The interactions between climate and energy policy considerations.
Source: Critchfield (1978)

winters on heating demand. Relating to the influence of climate on energy supply, factors that can be affected include: the research/exploration for sources; the siting of power plants; the method, materials, timing, and costs of construction of energy supply facilities; and the transportation of energy (e.g., routing of colliers and tankers, highway and railway maintenance). A knowledge of climatic factors is particularly important for assessing renewable energy sources such as solar energy, wind energy, and hydropower.

The role of climate in affecting energy demand and supply was discussed in detail in three contributions to a conference held in March 1980 (Bach *et al.*, 1980). McKay and Allsopp (1980) looked at many of the topics included in this chapter, including the effect of climate on energy demand and supply and the potential of renewable energy sources. In their discussion of energy demand, they point out that over one-third of the energy consumed in industrialized North America and about 50% of that consumed in Europe is used to overcome the direct or indirect consequences of climate. They show that, although the influence of climate on energy demand is most evident in the case of space heating, significant demands are also found in agriculture, transportation, and other outdoor industries. As far as energy demand is concerned, they state that, as a city grows, the unit-heating demand is reduced within and for a short extent downwind of the city. McKay and Allsopp find that a 1.6 °C temperature increase in Toronto (the actual increase observed during the past 100 years) would reduce space heating demands about 12%, although they point out that this could be partly offset by increased demands for air conditioning in the summer months.

In the agricultural sector, McKay and Allsopp state that in the production of corn, 50% of the energy use is for fertilizing, 25% for mechanical operations, 20% for corn drying, and 3.5% for pesticide use. Climate can influence each of these activities. Several examples are given of energy- and money-saving possibilities with proper use of climate information.

Page (1980) also addresses the interactions between climate and energy supply and demand and points out that, if the energy demand of human settlements is to be reduced, it is necessary to take into account the effect of urban form on building energy demand and on transportation and to examine the energy content of building materials. In this context, Page discusses the design of a city with materials and energy conservation. He shows that climatological aspects of energy conservation are very important, and there is considerable scope for the development of climatological awareness in the overall design of human habitats.

7.2   THE EFFECT OF CLIMATE ON SOLAR ENERGY CONVERSION SYSTEMS

As indicated in Chapter 6, there is a variety of technologies for the conversion of solar energy. These range from solar collectors on house roofs for space and water heating to solar power plants, with large arrays of mirrors concentrating solar energy to heat water and drive turbines to produce electricity. To some extent, the type of solar energy conversion system installed in an area depends on the climate. Solar thermal-electric conversion systems based on the 'power tower' concept, as described in Chapter 6, convert only direct radiation. Thus they are more suited to the dry desert-climate regions, where the proportion of direct radiation can be as high as 80%, than to the cloudier middle latitudes where the proportion of direct radiation is much smaller, especially in the winter months. Insolation assessment is therefore an important step in planning for the introduction of solar energy conversion systems. Insolation resource assessment programmes (e.g., Riches and Koomanoff, 1978; EC, 1980) have been established to collect, record, and archive climate data.

The use of solar energy for domestic space and water heating depends on a number of climatic factors, as has been discussed in detail by Jäger (1981). Not only does solar energy collection depend on climatic variables, but the heat losses in a house are also climatically sensitive. The energy losses through the walls, windows, and roof, etc., depend on the outside air temperature, the windspeed, and the effective temperature of the sky, which influence long-wave radiation losses. On the other hand, the heat gains of the house due to absorption of solar energy are influenced by such factors as the amount of cloudiness and the atmospheric clarity (e.g., the amount of water vapour in the air). The design of a solar house involves minimizing the heat losses in winter while maximizing the heat gains and also ensuring that the house is acceptably comfortable in summer through control of excessive summer heat gains by shading and other techniques. Obviously, it is necessary to understand and take into account the climatic factors. Moreover, for the design of a solar house, one needs information on the large-scale climate and on the local climatic conditions, influenced by shading, sheltering, topography, etc.

Jäger (1981) has emphasized the importance of making a balanced assessment of all climatic factors, when making an evaluation of the economic applicability of solar heating systems. For instance, he points out that, if one were to consider only the distribution of incoming solar radiation in Europe, one might draw the conclusion that solar heating would be more economically attractive in the southern locations, where the solar radiation availability is higher. However, the length of time over which heating is required is relatively short in the south because of the relatively higher temperatures associated with lower cloudiness, more sunshine, and less wind. A longer operating season in the less climatically favourable locations could give an installed solar space heating system greater economy.

Jäger (1981) has also shown for a number of European locations how the dimensioning and performance of solar space- and water-heating systems are related to climatic factors by analysing the performance of a solar space-heating system of the liquid type built on to a reference house. The collector area and storage volume of the solar heating system were determined such that approximately 50% of the annual space- and water-heating requirements would be covered by the system.

Figure 7.2 shows the performance results calculated by Jäger (1981) for the locations in Europe where hour-by-hour climate data are available. The heating

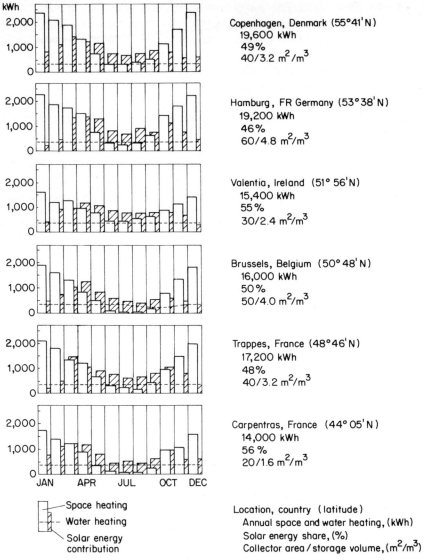

Figure 7.2   Performance of a solar heating system for a reference house calculated for six locations in Europe. Source: Jäger (1981)

requirements of the reference house reflect the variations of climate within Europe, with oceanic, continental, and Mediterranean influences. When the solar radiation availability and heating requirements coincide temporally, the required collector area and storage volume are obviously smaller than when the solar radiation is available at times when the heating requirements are low. Thus, although Hamburg and Copenhagen have similar heating requirements, the reference house in Hamburg requires a larger collector area, because the solar radiation is not so well correlated there with the heating needs. Figure 7.2 shows that in Ireland a relatively small collector area and storage volume give a good solar system performance. Jäger attributes this result to the influence on Irish climate of the ocean, which warms and cools much more slowly than the land. Thus, the winter heating requirements are comparatively low and during summer heating requirements still occur because the average daily temperature drops below 15 °C. Both factors contribute to a good solar space-heating system performance.

7.3 THE EFFECT OF CLIMATE ON WIND ENERGY CONVERSION SYSTEMS

A fraction of the solar radiation incident on the earth is converted by the atmosphere into the kinetic energy of the wind. This natural conversion provides a form of mechanical energy that has long been used by mankind, for example in sailing ships and windmills. The potential uses of wind power as an energy source have recently been recognized and, as outlined in Chapter 6, it is proposed to use wind turbines to generate electricity.

The distribution of wind energy varies markedly in space and time; thus, a knowledge of these distributions is necessary in order to select and site wind energy conversion systems effectively. As listed by Hardy and Walton (1977), several meteorological properties of any region are needed for wind-energy conversion systems studies:

— areas of smoothly accelerated or enhanced winds;
— locations of flow-separation zones;
— mean hourly wind velocity distributions;
— characteristics of local turbulence and gusts;
— occurrences of extreme winds and calms;
— vertical profiles of the wind velocity as a function of atmospheric stability;
— frequency of severe thunderstorms, lightning, hail, icing, tornadoes, or hurricanes;
— presence of salt spray or blowing dust.

However, data of these types rarely exist in many areas that are considered most appropriate for wind energy systems development. For an evaluation of the continental-scale wind energy resource, wind observations from the established meteorological observation net are available, although many of them are made at airports, which are generally not located in the windier areas. Traditional wind data are also very difficult to interpret in mountainous, hilly, and coastal regions.

Figure 7.3 shows the scales of interest for the three aspects of wind energy development: initial evaluations; site selection and assessment; and machine design and performance. The initial evaluations can generally be made on the basis of the standard meteorological observation net. Data for site selection and

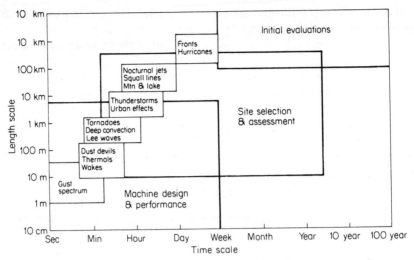

Figure 7.3   The time and space scales important to wind energy development.
Source: Hardy and Walton (1977)

assessment are more difficult to obtain because the density of measurements at scales less than 100 km is so poor, especially in mountainous areas.

Wind variations over long periods of time must also be considered in planning systems. Hardy and Walton (1977) report that long-term (10-year) variations in wind energy conversion systems output were recently estimated from standard observational data from 15 locations in the United States. Power output was estimated by the hour over the 10-year period at each site and averaged. They found that, in addition to expected intra-annual variations, significant year-to-year trends also occurred. Interannual variations of about 25% of the long-term mean were estimated at most locations. Year-to-year variations increased the difference between the minimum and maximum monthly energy production to 35–65% of the long-term mean production. However, they also estimated that minimum energy costs could be achieved at all sites with only a modest amount of energy storage to buffer changes in output.

The collection of wind power at levels near 100 MW represents a very significant undertaking. The system size must be planned, turbines sited, and array output estimated over weeks, months, and years (Hardy and Walton, 1977). Since existing meteorological stations cannot provide all the necessary data, other approaches are needed. For mountainous areas, for example, they would involve:

— collection of meteorological observations from multiple sources;
— use of a numerical model to simulate the windfield over terrain;
— statistical analysis of regional wind velocity patterns;
— coordinated application of the field measurement, statistical, and numerical modelling efforts.

A study incorporating these aspects has been conducted, for instance, for the island of Oahu, Hawaii (Hardy, 1977a and b).

The availability of renewable energy can be enhanced if the combined use of wind and solar energy is considered. Kahn (1979), for example, has shown the results of a study of joint wind/solar availability for a single site in Texas. At this location, the wind energy available at two heights was considered, but he found that the larger amount of wind energy available higher up produced excess power that was only available in the spring. A smoother aggregate of solar and wind energy was available using wind energy conversion at a lower height. Kahn points out that it is difficult to generalize from such results and that it is also necessary to look at smoothing effects, such as geographical dispersion. The relation between aggregate energy availability and geographical dispersion has been investigated more extensively for wind energy conversion than for solar energy conversion (Kahn, 1979). For example, Kahn cites a study for West Germany showing that at one site there was no wind energy output for 1,500 hours per year. When three sites were considered, there was no output for several hundred hours per year, and with 12 sites there was no output for less than 20 hours.

7.4   THE EFFECT OF CLIMATE ON HYDROPOWER SOURCES

Just as climate information is required for assessing solar- and wind-energy resources, requirements exist for climate data to evaluate hydropower resources. In particular, local and regional hydrological characteristics must be determined, with emphasis on the means of precipitation and evaporation and their variability. The requirements can be just as complex as those outlined above for other renewable energy sources, since the precipitation characteristics of a watershed may involve time series of snow accumulation and ablation and these plus evaporation will be influenced by prevailing temperatures, humidities, and wind characteristics. Thus, not only data but also relevant models are required to assess the potential energy source. Models are available that cover time scales ranging from hours to months. They can simulate run-off with an explained variance of over 95% using very few parameters (Wigley, personal communication). Similarly the long-term variability must be assessed. Droughts, for example, have been noted to affect the hydropower supply significantly: the drought in the northeast US during the period 1961–1966 reduced flow rates of rivers and reservoir levels; New York City reservoir levels, for example, were reduced to 40% of their capacity in 1965 (Beltzner, 1976).

McKay and Allsopp (1980) have discussed the effect of climate on hydropower supplies. They point out that storage must be provided to overcome climatic variability. The power generation of the Niagara river system in 1964 was reduced 26% by drought conditions. The drought in California in 1976–1977 consisted of 65% of normal precipitation in 1976 and only 45% in 1977. Hydroelectricity generation (20% of total on average) was reduced from about 33,000 million to 16,000 million kWh in 1976 and 13,000 million kWh in 1977 (McKay and Allsopp, 1980). Quirk and Moriarty (1980) have given further descriptions of the effect of the California drought. They state that the operating expenses of the California utility (Pacific Gas and Electric) in 1977 were about 30% greater than the 1976 operating expenses of about $1,275 million. Fifty million barrels of oil had to be burned statewide to make up the energy deficit. In addition, agricultural users were forced during 1977 to pump up more ground water. It is estimated that the extra energy associated with the additional pumping in California was approximately 1,000 million kWh.

7.5   THE EFFECT OF CLIMATE ON ENERGY DEMAND AND SUPPLY

The effect of climate on energy supply and demand has received attention recently because of the observed effect of cold winters, particularly in the eastern US. For example, Quirk (1979) points out that the increasing demand for energy and the dependence on foreign countries for supply has made the US energy system increasingly vulnerable to disruption. Events during the cold winter of 1976–1977 illustrate this point. Use of heating oil was 16% greater than normal and fuel shortages, particularly of natural gas, caused laying off of workers temporarily, closing schools, etc. The relevance of climatic research is illustrated by the fact that some of the economic disruption of that cold winter could have been avoided if adequate seasonal and monthly climate forecasts had been available. For example, as Quirk (1979) has pointed out, November was very cold and over 12% of the nation's stored gas was then depleted, whereas this reserve is not usually used in November. In December some gas suppliers continued to serve large gas customers who could have switched to oil. The early switchover to oil would have ameliorated the subsequent shortage, but would only have been justifiable if a reliable forecast that the cold would continue were available. Likewise, closing schools and factories in February could have been avoided if it had been known that March was going to be unusually warm. Other climatic effects included rivers freezing, which prevented delivery of coal, and the drought in the western US, which affected the hydroelectricity supply.

The increasing use of air conditioning and heating has increased the sensitivity of the energy demand to temperature changes. McQuigg (1975) showed that there was a noticeable increase in electricity demand as a function of increasing temperature because of the use of air conditioning.

Climate-related heating demand can be evaluated using 'heating degree-days'. A degree-day calculation requires that a base temperature be defined, usually taken to be 18 °C. For a day upon which heating is required, the number of heating degree-days equals the difference between the average temperature on that day and the base temperature. For example, if the average temperature on a particular day is 12 °C, the number of heating degree-days is 6. The number of heating degree-days can be added for the whole heating season and used as a guide to the annual heating requirement. The degree-day total for a number of days is generally proportional to the total heating load for that period. Also, the relation between degree-days and fuel consumption is usually assumed to be linear, i.e., if the number of degree-days is doubled, the fuel consumption should be doubled. Tabulated degree-day values are available for a large number of stations and reflect the long-term average situation.

Mitchell *et al.* (1973) have computed the seasonal total heating degree-days for each state of the US for each of 42 heating seasons from 1931–1932 to 1972–1973. This study allows one to estimate the influences of climate on heating-fuel demand. Results showed that in 1 year out of 100 one should expect the national total demand for heating fuel to exceed the long-term average demand (for constant economy) by as much as 10%. Similarly, the demand can be expected to exceed the average by at least 3.6% on an average of one heating season in five. However, when one part of the continent is colder than average it is not unusual for other parts to be warmer than average, thus the use of national averages reduces the observed variability. The probable extreme deviations of heating demand are larger when regions are considered, especially in the southern and Pacific states. For example, for the south Atlantic states, in 1 year out of 100 one should expect a total demand for heating fuel to exceed its long-term average demand by 20.4%. To the extent that fuels are not readily distributed from one part of the country to another, the results of Mitchell *et al.* are of great significance.

The time series of seasonal totals of heating degree-days for the US show that the greatest accumulation occurred over the United States during the winter of 1935–1936. Figure 7.4 shows an estimate of the per cent increase in heating fuel

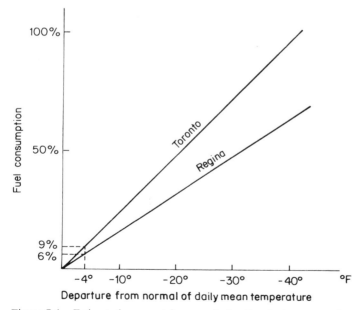

Figure 7.4   Estimated per cent increase in heating fuel consumption by temperature for Toronto and Regina for the period January, February, and March. Source: Beltzner (1976)

consumption as a function of temperature for Toronto and Regina for the winter season. On the basis of this figure it has been estimated that large areas of the northern United States and Canada would have had an increased fuel consumption of 50% or more in February 1936.

Figure 7.5 shows the October–March US national population weighted heating degree-days, using a base temperature of 65 °F (18.3 °C) for the period 1897–1898 to 1978–1979 calculated by Diaz and Quayle (1980). This figure shows that the heating degree-days were highest during the winter of 1976–1977 and in the subsequent two winters. In the long term the period between about 1920 and 1955 was marked by warmer conditions than the preceding and subsequent years, although individual years had higher than average numbers of heating degree-days.

O'Brien (1970) calculated degree-days for Melbourne, Australia and found that a decrease in mean outdoor temperature of 0.5 °C or 1.0 °C gave increases in heating requirements of 9.6% and 19.6% respectively. Such increases indicate that temperature variations could affect future energy demand significantly.

Manley (1957) has used a variant of the degree-day method, namely a method of degree-months, to examine the variations of the heating-season (Septem-

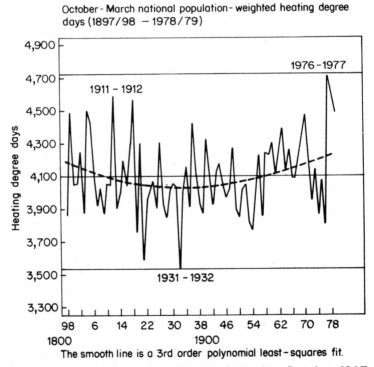

Figure 7.5 Total October–March heating degree-days (based on 65 °F base temperature) for the United States weighted by population. Source: Diaz and Quayle (1980)

ber–May) energy requirement since 1700 in the United Kingdom. He found that the seasonal requirement over the coldest decade exceeded that for the warmest decade by about 25%. Further, the requirement for the coldest individual heating season exceeded the average requirement by 36% and that for the warmest heating season by 85%.

Karl and Quayle (1981) have examined the relation between population-weighted cooling degree-days and electrical energy consumption during the hot and dry summer of 1980 in the southern portions of the United States. The cooling degree data were obtained from the US National Climatic Center. Monthly data of national electric utility sales beginning 1971 were obtained from the *Monthly Energy Review* (US Department of Energy, 1975–1980). Record electrical energy usage was reported during the 1980 heat wave, but, as Figure 7.6a shows, the national cooling degree total during the cooling season (May–October) was well below record levels (Karl and Quayle, 1981). The authors calculated that, since 1895, 13 out of 86 years had a greater number of cooling degree-days than 1980. The cooling degree total was mostly near or much below the long-term mean during the 1960s and 1970s. The 1970 population weights were used throughout the entire period to demonstrate how past weather conditions would influence contemporary conditions if they were to recur.

Karl and Quayle found a high correlation between cooling degree-days and electricity consumption. They computed that, at 1980 consumption levels, five national cooling degree-days (population-weighted) represented $\sim 10^9$ kWh of weather-sensitive electricity demand. The relation is very strong for residential and commercial customers (Figure 7.6b and c), which together account for $\sim 56\%$ of all sales. The authors point out that the apparent lag in consumption (peaking one month after cooling degree-days) is primarily the result of increased lighting requirements as the daylight duration shortens. The industrial sector (Figure 7.6d) is poorly correlated with cooling degree-days as a result of summer shutdowns.

Karl and Quayle (1981) discussed the implications of their results for models relating climate to energy consumption. They pointed out that, if power production values are used (rather than sales), weekly statistics are possible and time resolution is very good, but only total production is known, not individual consumption values for residential, commercial, industrial, and other users. However, it is clear in Figure 7.6 that the industrial sector must be treated differently from the other sectors. Further complications listed by Karl and Quayle are: holidays and weekends; length of the day; other weather-sensitive uses such as irrigation; variable line losses; miscellaneous sales and losses; uncertain growth projections; conservation measures; and separation of grow-related trends and climate-related trends.

7.6 WEATHER AND CLIMATE FORECASTING

A further interaction between climate and energy demand/supply is in the area of forecasting. Here it is difficult to draw the line between weather and climate, inasmuch as climate data are necessary for forecasting extreme events and the

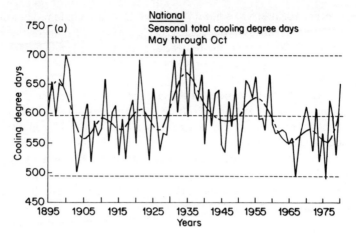

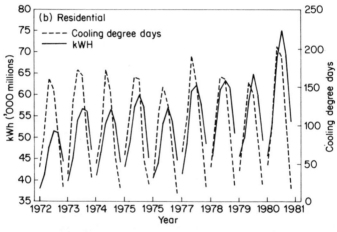

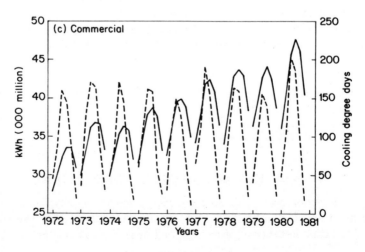

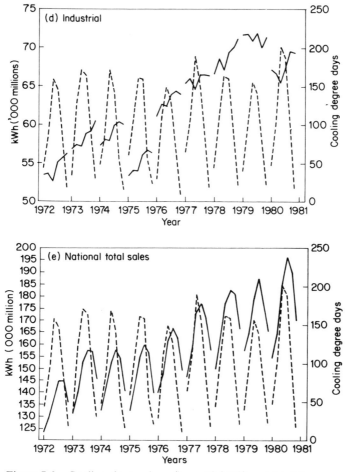

Figure 7.6   Cooling degree-days (base 18.3 °C) weighted by population (May–October) compared to electric energy sales. Source: Karl and Quayle (1981)

development of predictive capabilities for day-to-day weather and longer-term climate are interlinked. As pointed out by Suomi (1975), weather forecasts for one to seven days would provide the management of utilities with advanced warnings of several local weather conditions. Suomi suggests that advanced knowledge of the severe icing conditions in the northeastern United States in December 1973 would have enabled utility management to take preparatory actions reducing the level and duration of the lack of electricity supply.

More accurate medium- to long-range forecasts would also be of value, especially through accurate forecasts of periods of unusually low or high temperatures or premature seasonal changes. Suomi (1975) indicates that spot shortages of petroleum products and coal experienced in the United States during

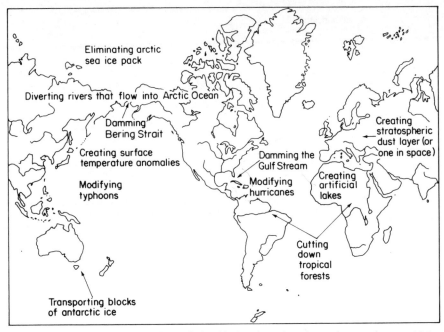

Figure 7.7 Schematic illustration of the kinds of engineering schemes that could be proposed to modify or control the climate. Source: Kellogg and Schneider (1974)

the 1972–1973 and 1973–1974 winter seasons could have been partially averted by improved weather forecasting. Salstein (1978) has shown that small temperature-forecast errors cause extra millions of cubic metres of natural gas to be required by moderate-sized cities to protect against optimistic errors in cold weather periods. All of these examples and that of Quirk (1979) cited in the last section illustrate the need for climate data, models, and services to aid in forecasting energy demand and supply.

7.7 INTENTIONAL CLIMATE MODIFICATION

Some study has been made in the last 30 years of the possibilities of weather modification. There are several areas in which weather modification could be used to benefit energy systems; they have been discussed, for example, by Suomi (1975). It is reported that increased precipitation from warm or cold clouds has already provided hydroelectric power and water for removal of waste heat and other products from power plants. Research is going on into modifying severe winter storms, hailstorms, and hurricanes; advances in treating any severe storms would be of value in terms of the disruption they cause to energy systems. Augmentation of winter snowpack is an area with obvious benefits for the production of hydroelectric power. A low stratus deck at night, produced by locally augmenting humidity could serve as a heat shield decreasing energy requirements on cold nights. On the other hand, cloud reduction would be helpful for solar energy systems. If clouds could be manipulated systematically to alter radiation and thus surface temperatures, numerous opportunities would arise to increase the yield of solar energy conversion systems.

There still exists considerable uncertainty about the consequences of intentional modification of the large complex weather phenomena. This is especially true of the downwind effects from large cloud-seeding programmes.

In addition to intentional weather modification, several schemes have been proposed for intentional climate modification; they are summarized in Figure 7.7. These schemes, which have been described (but not devised) by Kellogg and Schneider (1974), have been proposed for the purposes of cooling or warming vast regions of the earth, changing patterns of rainfall, protecting humans or property from damaging storms, etc. They could be used to improve the current climate (for some) or to offset a predicted deterioration of climate (for some). However, as Kellogg and Schneider (1974) have pointed out, it would be dangerous to pursue any large-scale operational climate control schemes until we can predict their long-term effects with some acceptable assurance. We are not able to do this now and it will be some time before we can; perhaps it will not be possible at all.

7.8   SOME CONCLUDING REMARKS

As illustrated in Figure 7.1, there are a large number of interactions between the climate system and the energy systems. While most of this book concentrates on the effects of energy systems on climate, topics covered in this chapter indicate that there are significant interactions in the other direction too. In particular, it can be seen that large amounts of many kinds of climate data are required to assess certain energy supply systems and when these data are unavailable there is a requirement for models to interpolate or reconstruct relevant information. Likewise, we have seen the value of climatological data and weather and climate predictive capabilities for forecasting of energy demand. It is also clear that we need an adequate theory of the physical basis of climate and climatic change and an ability to predict future climate variability.

REFERENCES

Beltzner, K. (ed.) (1976). *Living with Climatic Change*. Science Council of Canada, Ottawa, Canada.

Critchfield, H. J. (1978). Climatology in a comprehensive energy policy. American Meteorological Society, Conference on Climate and Energy: Climatological Aspects and Industrial Operations, 8–12 May, Asheville, NC.

Diaz, H. F., and R. G. Quayle (1980). An analysis of the recent extreme winters in the contiguous United States. *Mon. Wea. Rev.*, **108**, 687–699.

EC (1980). Atlas über die Sonnenstrahlung in Europa. Band 1: Globalstrahlung auf horizontale Flächen. W. Grösschen-Verlag, F.R. Germany.

Hardy, D. M. (1977a). Numerical and measurement methods of wind energy assessment. *Proc. Third Biennial Conference on Wind Energy Conversion Systems*. Washington, DC, 19–21 September 1977.

214

Hardy, D. M. (1977b). Wind studies in complex terrain. *Proc. American Wind Energy Association Conference and Exposition*. Boulder, Colo., 11–14 May 1977.

Hardy, D. M., and J. J. Walton (1977). Wind energy assessment. Miami International Conference on Alternative Energy Sources, 5–7 December 1977, Miami Beach, Fla.

Jäger, F. (1981). *Solar Energy Applications in Houses*. Pergamon Press, Oxford, England.

Kahn, E. (1979). The compatibility of wind and solar technology with conventional energy systems. *Ann. Rev. Energy*, **4**, 313–352.

Karl, T. R., and R. G. Quayle (1981). The 1980 summer heat wave and drought in historical perspective. *Mon. Wea. Rev.*, **109**, 2055–2073.

Kellogg, W. W., and S. H. Schneider (1974). Climate stabilization: for better or for worse. *Science*, **186**, 1163–1172.

Manley, G. (1957). Climatic fluctuations and fuel requirements. *Scottish Geogr. Mag.*, **73**, 19–28.

McKay, G. A., and T. Allsopp (1980). The role of climate in affecting energy demand and supply. In, W. Bach, J. Pankrath, and J. Williams (eds.), *Interactions of Energy and Climate*. Reidel, Dordrecht, Holland.

McQuigg, J. D. (1975). Economic impacts of weather variability. Atmospheric Science Department, University of Missouri, Columbia, Missouri.

Mitchell, J. M., R. E. Felch, D. L. Gilman, F. T. Quinlan, and R. M. Rotty (1973). Variability of seasonal total heating fuel demand in the United States. National Oceanic and Atmospheric Administration, Washington, DC.

O'Brien, L. F. (1970). Heating degree days for some Australian cities. *Australian Refrigeration, Air Conditioning and Heating Journal*, **24**, 36–37.

Page, J. K. (1980). Climate considerations and energy conservation. In, W. Bach, J. Pankrath, and J. Williams (eds.), *Interactions of Energy and Climate*. Reidel, Dordrecht, Holland.

Quirk, W. M. (1979). Prospects for monthly forecasts with 3-D global weather models. Paper presented at Fourth Conference on Numerical Weather Prediction. 29 Oct.–1 Nov., Silver Spring, Maryland, Amer. Meteor. Soc.

Quirk, W. J., and J. E. Moriarty (1980). Prospects for using improved climatic information to better manage energy systems. In, W. Bach, J. Pankrath, and J. Williams (eds.), *Interactions of Energy and Climate*. Reidel, Dordrecht, Holland.

Riches, M. R., and F. A. Koomanoff (1978). The national isolation resource assessment program: A status report. American Meteorological Society Conference on Climate and Energy: Climatological Aspects and Industrial Operations, 8–12 May 1978, Asheville, NC.

Salstein, D. (1978). Limitations to the accuracy of energy resource allocation based on weather prediction. American Meteorological Society Conference on Climate and Energy: Climatological Aspects and Industrial Operations, 8–12 May 1978, Asheville, NC.

Suomi, V. E. (1975). Atmospheric research for the nation's energy program. *Bull. Amer. Meteor. Soc.*, **56**, 1060–1068.

WCC (1979). *Proceedings of the World Climate Conference*. WMO Publication No. 537, World Meteorological Organization, Geneva.

CHAPTER 8

Summary and Conclusions

8.1 ENERGY USE IN THE FUTURE

In 1975 the global primary commercial energy consumption was about 8 TWyr/yr. Most of this supply came from fossil fuels, with oil supplying almost half of the global primary energy. The consumption of energy is very unevenly distributed throughout the world. Although the average global primary energy consumption is about 2 kWyr/yr *per capita*, roughly 70% of the world's population lives with less than the average and most of this 70% with only 0.2 kWyr/yr of commercial energy consumption. By the year 2030, it is projected that the world's population will have reached a steady total of about 8,000 million, compared with about 4,000 million now. If the *per capita* energy consumption remains the same, the global energy consumption would be 16 TWyr/yr in 2030. On the other hand, if the *per capita* consumption increases, as it is likely to do if an attempt is made to even out the distribution, then a 5 kWyr/yr *per capita* demand would give a total primary energy demand in 2030 of 40 TWyr/yr. Thus, plausible projections suggest that the global primary energy consumption will be between two and five times larger than that of the present day in the year 2030.

The IIASA Energy Systems Program (1981) considered two scenarios for global energy supply and demand with total global primary energy consumptions in the year 2030 of 35.65 TWyr/yr and 22.39 TWyr/yr. More than 60% of the energy supply in the year 2030 will still come from fossil fuels, according to the IIASA study. However, nuclear and solar power will have a larger share than at present, with the transition from the use of fossil fuels to a reliance on renewable energy sources occurring after 2030. During the next 100 years, fossil fuels, nuclear power, and renewable energy sources will supply energy. This book has investigated the possible ways these energy supply sources could have effects on climate.

8.2 THE CLIMATE SYSTEM

The climate system is very complex. It consists of the atmosphere, the oceans, the ice and snow cover, the land surface, and the terrestrial biota. These components interact with one another through processes such as evaporation and wind stress. For example, the atmosphere exerts wind stress on the ocean surface; the ocean surface water evaporates into the atmosphere, the land surface exchanges sensible heat with the atmosphere, and so on. Because of these interactions, changes in one part of the climate system can lead to changes in another part. For instance, changes in the ocean surface temperature in a particular region can lead to changes in the atmospheric storm tracks, leading to changes in temperature, rainfall, and other climatic variables downstream. Thus climatic variations can be a result of feedbacks between the variables within the air, water, and ice. One feedback loop that is often described is the ice–albedo feedback loop, in which an increase in the amount of solar radiation leads to a decrease of ice cover, giving a lower reflectivity, so that more radiation can be absorbed and more ice melts. This is a positive feedback loop. There are also negative feedbacks.

As a result of natural forcing mechanisms, climate has changed and is changing on all time and space scales. On the global scale the climate is known to have changed on a very long (geological) time scale. About 20,000 years ago the earth was experiencing a glacial period with large continental ice caps in North America and Europe. On a shorter time scale the Northern Hemisphere was on the average somewhat cooler during the first half of the 19th century than at present, a warming occurred until 1940, a cooling followed until the middle of the 1960s. Since then there is some evidence of a warming trend.

Present understanding of the complexities of the climate system and knowledge of the changes that have occurred in the past indicate that the climate can be sensitive to natural and anthropogenic changes in climatic boundary conditions. The question then arises: Will energy systems also perturb the climate system? We have found that such changes could occur because of waste heat release, additions of CO_2 and other gases and particles, or because of changes in characteristics of the earth's surface such as wetness, roughness, or reflectivity.

Energy systems could affect climate on the local, regional, or global scale. We have already observed that, on the local scale, climatic variables (e.g., temperature and cloudiness) can be changed in the vicinity of a power plant. On the regional scale, urban-industrial areas influence precipitation and cloudiness, not only overhead, but also several tens of kilometres downstream. On the global scale there is no evidence that man's activities have influenced the climate.

One reason for concern about the potential effects of energy systems is the possible undesirability and irreversibility of climatic changes. Thus, we want to be able to predict the effects of energy systems before embarking on an energy strategy that has unwanted environmental consequences.

Unfortunately, it is not possible at present to predict with any reliability the climatic consequences of particular energy strategies or scenarios. It is, however, possible to investigate the sensitivity of the climate system to certain imposed changes and thus to compare the magnitude of the influences of different factors. The main tool for making such sensitivity analyses is the numerical climate model, although climatic analogues can also be useful in sensitivity analyses.

8.3   CARBON DIOXIDE

At present the effect that is thought to be potentially the most serious arises from the release of CO_2 by fossil-fuel combustion. Since 1958 accurate measurements of the amount of CO_2 in the atmosphere have been made. The annual average concentration of CO_2 measured at Mauna Loa has increased from just over 315 ppm in 1958 to 338 ppm in 1980. The increase in atmospheric CO_2 concentration is a global phenomenon. It is believed that part or all of this increase is due to the release of CO_2 into the atmosphere by burning fossil fuels (coal, oil, and gas). It has been argued recently that the release of CO_2 as a result of deforestation and soil deterioration, especially in the tropics, has also contributed to the observed CO_2 increase. Concern arises because of the possibility that substantial increases in the CO_2 concentration would lead to possibly irreversible and undesirable climatic changes. It thus appears necessary to find out whether the CO_2 concentration can be expected to continue to increase, and this involves understanding the reasons for the presently observed increase. It is also necessary to understand the natural carbon cycle and to be able to predict the climatic consequences of CO_2 increases.

The four reservoirs of carbon are the atmosphere, the ocean, the biota, and sediments. There are two possible sinks for the CO_2 released into the atmosphere by fossil-fuel combustion: the ocean and land vegetation. The capacity of the oceans to store carbon is tremendous, but the transfer to the deep ocean is slow. Terrestrial vegetation can act either as a source or as a sink of carbon. The main storage is in the forests; it has been claimed that clearing forests is a source of CO_2.

The carbon cycle is very complex. There are a number of reservoirs of carbon with chemical, physical, and biological processes linking them. Numerical models of the carbon cycle are necessary so that predictions of the future concentrations of atmospheric CO_2 can be made. Although there are many uncertainties in the models of the carbon cycle, it is possible to make predictions of the future atmospheric CO_2 concentration, particularly for a period of 20–30 years from now. There is a general consensus among oceanographers on how much fossil-fuel CO_2 the oceans can remove from the atmosphere. The role of the biosphere is more uncertain. Deforestation, especially in the tropics, has been a source of atmospheric CO_2. Regrowth of forests, particularly in temperate latitudes, and enhanced growth due to atmospheric CO_2 increases have been acting as a sink for atmospheric CO_2. Theoretical and observational studies of the ocean and terrestrial vegetation suggest that the biota cannot at present be a net source of CO_2 as large as the fossil-fuel source.

Many studies have made projections of the future use of fossil fuels and the consequent atmospheric CO_2 increase. They generally make a prediction of the energy demand in the future and adopt assumptions about how much of this demand will be satisfied by fossil fuels. Another approach is to investigate how much fossil fuel would have to be burned to give a certain CO_2 concentration by a given time and to ask where the fossil fuels would come from. This approach finds that CO_2 could become a problem mainly as a result of long-term large-scale development of coal resources by a small number of countries. However, constraints on international coal trade and the technical feasibility of large-scale coal exploitation could prevent CO_2 from reaching threatening levels in the coming century. Nevertheless, most projections of the future levels of atmospheric CO_2, generally based on models of the carbon cycle and assumptions about energy supply and demand, envisage a continued increase in the amount of atmospheric CO_2 due to burning fossil fuels. Many projections suggest that a doubling of the preindustrial CO_2 concentration will occur within the next 100 years.

8.4 THE EFFECTS OF CO_2 ON CLIMATE

Many studies made with climate models of varying complexity have shown that increased atmospheric CO_2 produces a warming of the earth's surface and of the lower atmosphere. This warming is due to the fact that CO_2 is a good absorber/emitter of long-wave radiation. The surface warming is therefore caused by the increased downward emission of long-wave radiation from the CO_2 in the lower atmosphere.

Calculations show that doubling the atmospheric CO_2 concentration would lead to a net heating of the lower atmosphere, oceans, and land by a global average of about 4 W m^{-2}. There is relatively high confidence that this net heating value has been estimated correctly to within $\pm 25\%$. Greater uncertainties arise in estimates of the change of global mean surface temperature resulting from the change in heating rate. The present best estimate is that the global mean surface temperature change would be 1.6–4.5 K for a doubling of the atmospheric CO_2 concentration.

The response of the climate system to an increased CO_2 concentration cannot be expected to be uniform in space or time. It is necessary to know not only what the change in global average temperature would be, but also how meteorological variables (rainfall, temperature, wind speed, etc.) would change at different times of the year and in particular regions. One major effort to answer such questions is being made with the use of numerical models of the general circulation of the atmosphere or atmosphere–ocean system. At present such models have many shortcomings, including poor treatment of clouds, precipitation, and orographic effects. Consequently, the existing models are not reliable in their predictions of regional climatic changes due to changes in the CO_2 concentration.

The models that have been used so far to investigate the effects of increased CO_2 have either not treated the oceans specifically or have greatly simplified the ocean and its interactions with the atmosphere. A change in the atmospheric circulation due to increased CO_2 would certainly lead to changes in the ocean circulation and further feedbacks on the atmosphere. Thus, there is a need to consider the entire ocean circulation in simulations of CO_2 effects. A further effect of the ocean on the CO_2 effects is that it introduces a time lag in the climatic response to forcing. A delay of up to several decades in the surface temperature response could occur owing to the heat capacity of the oceans. That is, the globally averaged surface temperature may reach its equilibrium value several (or many) years after the doubling of the CO_2 concentration. Considerable attention needs to be given to the time and space variations of thermal inertia and to the transient (rather than equilibrium) response of the climatic system to a CO_2 increase.

Other approaches to predicting regional climatic changes have also been taken. One possibility is to study climatic evidence from periods when the earth was warmer than it is now, such as the Altithermal period 4,000 to 8,000 years ago. One cannot accept the results of such studies as a literal representation of what might occur if the earth becomes warm again, since the causes of the warming could have been quite different from those of the potential warming due to increased atmospheric CO_2 concentration. A similar approach has been taken in recent studies that have used instrumental observations of temperature, precipitation, and pressure during the last century as a basis for discussing the response of the climate system to warming. Although these studies cannot provide predictions of the changes to be expected due to an increase of the atmospheric CO_2 concentration, they should be useful in guiding the development of scenarios of potential changes. The approach illustrates clearly that large coherent anomalies are a basic response to climatic forcing.

One of the components of the climate system which could be influenced particularly by the changes induced by a CO_2 increase is the ice and snow cover. The various forms of ice and snow would each respond differently to a warming. The effects of a warming on the ice sheets of Greenland and Antarctica are likely to be complex and on a much longer time scale than considered for snow and pack-ice. Melting of the ice sheets would give a rise of the world sea-level but this is not considered to be a serious threat in the near future.

A number of measures have been proposed for preventing or removing a 'CO_2 problem'. They involve reducing the use of fossil fuels, removing CO_2 from the gases in power-plant stacks, removing CO_2 from the atmosphere or ocean, and adopting methods to cause a global cooling to counteract the potential global warming due to CO_2. Results of studies made with models of the carbon cycle suggest that reducing CO_2 emissions from fossil-fuel combustion could have a considerable effect on the atmospheric CO_2 concentration. Varying assumptions regarding the level of fossil-fuel use were shown in one study to give an atmospheric CO_2 concentration in the year 2100 ranging between 1,500 ppm (by volume) and less than 400 ppm (by volume). Several technologies are available to reduce emissions of gases and particles from power plants; however, they generally add to the cost and reduce the efficiency of the power plant. Technologies are also available to remove CO_2 from the atmosphere and ocean, but for most of the processes more CO_2 would be generated than recovered if a coal-burning power plant were used to supply the required electrical power. Options for disposing of captured CO_2 include disposal in the deep ocean, burial in oil and gas fields, fixation in natural clay, or disposal in outer space.

A further approach to CO_2 control is to recover CO_2 and reuse or recycle it.

The most discussed use of recovered carbon is its conversion into liquid or gaseous fuels using a non-fossil energy source. Most of the suggestions for manipulating climatic boundary conditions to counteract a CO_2-induced warming are speculative and the engineering feasibility has not been thought out. Since it is impossible to predict at the present time what the effects of any large-scale modification scheme would be, it would be irresponsible to tamper with the climate system deliberately.

If the atmospheric CO_2 concentration continues to increase and climatic changes either occur or can be reliably predicted, there are a number of potential responses. In any event, the climatic changes would not be uniform over the globe and questions of equity arise because some regions may be seen to benefit (e.g., more rainfall in presently arid areas), while others may be seen to suffer (e.g., less rainfall or higher temperatures in hot, arid areas). The CO_2 question is an interdisciplinary issue that needs international cooperation.

8.5   WASTE HEAT RELEASE

When averaged over the globe, the amount of heat released by human activities is a small fraction of the solar radiation absorbed at the earth's surface. The global average energy use is 10^{-4} of the solar energy absorbed at the earth's surface. However, at individual places on the earth's surface the heat release due to human activities is of the same order of magnitude as or larger than the absorbed solar energy. Assuming the unlikely values of a global population of 20,000 million with a *per capita* consumption of 20 kWyr/yr, we find that the global average heat release would be about 0.5% of the global average solar radiation absorbed at the earth's surface, giving a global average surface temperature increase of around 1 K.

Since, however, waste heat release is and will be concentrated in certain regions, it is more important to look at the effects of such sources than at the effect of a globally distributed source.

There are several published reviews of the atmospheric effects of power plants on the local scale. The most frequently observed change is due to low stratus or fog formation in the plumes from cooling ponds and towers. There are only a few reports of precipitation enhancement in the vicinity of power plants, and temperature effects also appear to be negligible. The maximum amount of electrical power currently generated at a single power station is about 3 GW(e) and the atmospheric effects of current heat dissipation rates are not serious problems, especially beyond the confines of the power station, provided that efforts are made to design the facility such that downwash is eliminated, drift is minimized, and plume rise is maximized.

In contrast to the present dispersed sites for power generation, some proposals have been made to build more than one power plant at a site, referred to here as a 'power park'. Several recent studies have considered the effect of proposed 10–50 GW(e) power parks on climate. The results of model and analogue studies and comparisons with natural phenomena suggest that the principal effects of the release of large amounts of waste heat from power parks would be, on the local and regional scale, significant changes in cloudiness and precipitation with an increase in the probability of severe weather (e.g., thunderstorms and tornadoes).

In addition to considering the effect of waste heat release on local and regional climate, the possible effect on global climate of unrealistically large releases of waste heat has been examined. This amounts to considering whether large-scale effects could occur if heat releases in particular places were large enough to interfere with weather systems on the scale of cyclones and anticyclones and cause further anomalies downstream in areas where there is no heat release. It has already been observed that certain natural anomalies, for example large-scale ocean-surface-temperature anomalies, can have an influence on the atmospheric circulation. The potential effect of waste heat release has been evaluated by using models of the atmospheric circulation. The model studies involve comparing the simulated atmospheric circulation in a model integration without any added perturbation (a control case) with the circulation including a prescribed addition of waste heat. Because of model shortcomings one cannot predict how the atmosphere would respond to a particular heat input, but the model studies are a guide to the order of magnitude of any response.

Overall, the model simulations indicate that the input of large amounts of waste heat (of the order of 100 TWyr/yr) at particular locations would cause large, coherent changes in the atmosphere, not just over the areas of heat input but also elsewhere in the hemisphere. The response may vary according to the location, amount, and manner of heat input. With heat inputs of the order of magnitude expected during the next century no significant hemispheric response in the model was discovered, but large regional changes over the areas of heat input were observed. Thus, it would seem that waste heat release from an energy use of the order of 30–50 TWyr/yr would probably not affect the global climate system significantly, although local and regional climatic effects could occur.

8.6   THE EFFECTS OF PARTICLES AND GASES OTHER THAN CO_2

In addition to releasing CO_2 and waste heat into the atmosphere, energy systems can influence the atmospheric concentrations of other gases and particles. Combustion of fossil fuels releases particles directly into the atmosphere; particles are also formed from the gaseous products of combustion. The effects of particles on the earth–atmosphere radiation balance, and thus on such variables as global surface temperature, are extremely complex. They depend on characteristics of the particles, such as shape, size, and radiative properties. The location of the particles with regard to their height in the atmosphere and the radiative properties of the underlying surface are also important. Particles also can change horizontal and vertical temperature gradients. For instance, man-made particles in the lower atmosphere generally (all other factors being constant) warm the lower atmosphere and cool the earth's surface, thus changing the vertical stability. Particles also influence the reflectivity of clouds formed with the particles as cloud condensation nuclei. The entire set of radiative, thermodynamic, and dynamic effects of particles can be assessed ultimately only with the aid of suitably detailed climate models, which can predict cloudiness and consider cloud–radiation interactions. Particles can act as cloud condensation nuclei; there is evidence of increased cloudiness and precipitation as a result of increased atmospheric particle loading.

Sulphur dioxide is also added to the atmosphere by man's activities, including burning fossil fuels, and it can have climatic effects. It has been estimated that, if sulphate from sea spray is ignored, 60% of the atmospheric sulphur is anthropogenic. Calculations have shown that the scattering of radiation by sulphate aerosols is equivalent, when spread over the Northern Hemisphere, to a cooling of 0.03–0.06 K. However, more detailed studies, taking into account such factors as infrared radiation, cloudiness, and radiation absorption by sulphate-containing aerosols, are required before an evaluation of the climatic effects of regional concentrations of SO_2 particles can be made.

A further gas with potential climatic impacts is N_2O. It has been calculated that by the year 2000, the greenhouse effect due to increasing atmospheric N_2O from using fertilizers and burning fossil fuels may be about 40–50% of that due to increasing CO_2, while by 2050 it may be about 20–30% of the CO_2 effect. Other man-made trace gases also have a greenhouse effect and could cause a warming of the earth's surface if their concentration continues to increase. The gases N_2O, CH_4, and NH_3 from fertilizers and fossil fuels and the chlorofluorocarbon gases used in spray cans and as refrigerants have the most significant potential effect, which is, however, less than that calculated for CO_2 and water vapour. Further studies have shown that tropospheric ozone also increases owing to fossil-fuel sources of CO, NO, and CH_4. Tropospheric ozone also has a greenhouse effect. It is estimated that the combustion of fossil fuels is a source of CO, CH_4, and NO that is about 25% of the natural source. A continuation of the present growth rate of emission of these gases has been calculated to lead to doubling the tropospheric ozone concentration in the next century, giving a global average surface temperature increase of 0.9 K. It has been estimated that anthropogenic sources of trace gases other than CO_2 may contribute as much as 40% of the surface warming to the combined surface warming effects of CO_2 and these gases. It is therefore obvious that trace gases other than CO_2 could play a significant role in any potential climatic changes due to man's activities. There are, as has been indicated earlier in this chapter and in Chapter 3, many uncertainties regarding the effects of increased CO_2 on the climate. Nevertheless, sensitivity studies with a wide range of models have been carried out and an understanding at least of the types of potential effect is emerging.

In the case of the evaluation of the effects of man-made particles, the studies are much more preliminary and even the direction of potential global changes remains uncertain. The potential effects of trace gases also have been studied mostly with one-dimensional climate models. However, the predicted effects are essentially an enhancement of the CO_2 effects, so that the lack of depth of study is not as serious as it is in the case of particles.

8.7   EFFECTS OF CHANGES IN THE EARTH'S SURFACE CHARACTERISTICS

A number of technologies exist or have been proposed for the use of renewable (mostly solar) energy sources. Many of these technologies involve changing the earth's surface characteristics. For example, solar thermal electric conversion using the 'power tower' concept involves installing arrays of steerable mirrors on the ground. They would change the surface energy balance, surface roughness, and perhaps the surface wetness if the area were paved. Likewise, hydropower production could require flooding an area with consequent changes in surface characteristics. Biomass conversion could require large plantations of trees, or irrigation.

The question then arises of whether these changes in boundary conditions could be large enough to influence local, regional, or global climate. Since many of the solar energy conversion schemes are likely to be deployed only on a small scale in isolated regions, the effects are likely to be very localized and to be very dependent on local conditions.

For the effects of solar thermal electric conversion plants, rough estimates suggest that the effects on the surface energy balance would not lead to a change in the amount of heat added to the atmosphere but in the manner in which it is added. The influence of such energy-balance changes has been investigated using a two-dimensional mesoscale model of the atmosphere. It has been found that a hypothetical STEC plant in southern Spain, generating about 30 GW(e), increased local cloudiness and precipitation in summer simulations.

Photovoltaic cells could be deployed in a number of ways, ranging from arrays of cells on roofs of houses to modules of all arrays for a central power station. At present the use of photovoltaic conversion in a decentralized fashion appears to be the preferred option, in which case local climate effects would not differ in magnitude from those currently observed in urban areas.

Biomass can be converted to solid, liquid, and gaseous fuels from numerous sources and by many technologies. This diversity means that no simple evaluation of the local and regional climatic effects of biomass conversion can be made. Probably the largest effects could be expected to occur with forest plantation areas, since forests have a significant influence on the exchange of water between the earth's surface and the atmosphere and a large dynamic interaction with the atmosphere. Particle emissions to the atmosphere due to biomass conversion schemes would probably not contribute significantly to the presently estimated anthropogenic inputs to the atmosphere.

The main effect on local climate of wind energy conversion systems is likely to be on a microclimatological scale unless huge turbine arrays are considered, which seems unlikely. It has been suggested that there might be some slowing of the wind for a short distance downwind of an array of windmills, but that the wind would rapidly accelerate because of the downward transport of momentum from the stronger winds aloft.

Since most renewable energy conversion schemes that could be used on a large scale are still in the development stage, their deployment on a scale large enough potentially to influence global climate will, in any case, occur in the distant future. Moreover, the detailed physical characteristics of potential large-scale systems are not well defined. It is, therefore, unrealistic to examine potential climatic effects of specific scenarios for the large-scale use of these technologies. It is possible, however, to describe the general types of effect that could arise due to large-scale changes in surface energy balance, roughness, and hydrological characteristics (due, for instance, to large-scale use of STEC, photovoltaic conversion, or biomass plantations) or due to large-scale changes in ocean surface temperatures from OTEC systems. A number of model studies have indicated that large-scale decreases in surface albedo lead to increased cloudiness and rainfall over the perturbed area with further effects on the general atmospheric circulation. Large-scale roughness changes could influence the atmospheric circulation due to changes in the transfers of momentum, heat, and moisture in the boundary layer. Large-scale anomalies in surface wetness could lead to effects on the upstream and downstream atmospheric circulation. Model and observational studies indicate that large-scale ocean-surface-temperature anomalies, which could occur with deployment of ocean thermal electric conversion systems, can have a significant influence on the atmospheric circulation.

Since most renewable energy conversion will be on a small and decentralized scale with different locations being more suited to particular conversion systems (e.g., windy areas to wind energy conversion, sunny areas to direct solar energy conversion), it seems likely that the effects on climate will be at most very localized. The large-scale deployment of some systems, which would take place in any event in the distant future, could cause more extensive and thus more significant changes in the earth's surface boundary conditions and thereby have a larger climatic effect.

8.8   THE EFFECT OF CLIMATE ON ENERGY SUPPLY AND DEMAND

The effect of climate on energy supply and demand has not been studied in as much detail as the effects of energy systems on climate. However, recent increasing awareness of society's dependence on climate and vulnerability to climatic change suggests that the topic will be considered more seriously in the future.

There are many ways in which climate can affect the demand for and supply of energy. For example, climate can influence the research and exploration for energy sources, since, for instance, exploratory drilling for oil in the Gulf of Mexico involves climate considerations quite different from those for drilling on the north slope of Alaska. Selection of sites for power plants also requires climatic information. Transportation, transmission, and storage of energy can also be influenced by climate.

Information on climate and climatic variations is particularly important for evaluating renewable energy sources. To assess solar radiation availability for space and water heating, it is obvious that information on the amounts of incoming solar radiation is required. In addition, however, a considerable amount of other climatic information is required. This is particularly important because the size of the required solar collectors depends on the correlation in time between the solar radiation available and heating requirements. When solar radiation availability and heating requirements coincide temporally, the required collector area and store volume are obviously smaller than when solar radiation is available at times when the heating requirements are low.

Similarly, to select and site wind energy conversion systems, knowledge of several meteorological characteristics of potential sites is required. The distribution of wind energy varies markedly in space and time, and this must be studied in detail.

The effect of climate on energy supply and demand has received attention recently because of the observed influence of cold winters, particularly in the eastern US. During the cold winter of 1976–1977, the use of heating oil was 16% greater than normal; fuel shortages, particularly of natural gas, caused temporary layoffs of workers, closing schools, etc. The relevance of climatic research is illustrated by the fact that some of the economic disruption of that cold winter could have been avoided if adequate seasonal and monthly climate forecasts had been available.

A further potential interaction between energy systems and climate is in the area of intentional weather or climate modification. For instance, it has been reported that artificially enhanced precipitation has already provided hydroelectric power and water for removal of waste heat and other products from power plants. However, enormous uncertainty exists about the consequences of intentional modification and it would be irresponsible to begin large-scale modification programmes, particularly since the consequences cannot be predicted. An improved understanding of the entire climate system and models for the reliable prediction of the effects of man-made changes are required before the climate system could be modified for the benefit of energy systems.

8.9 CONCLUDING REMARKS

The potential effects of CO_2 from fossil-fuel burning are believed to present the largest threat in the next 100 years. The atmospheric CO_2 concentration is observed already to be increasing globally, although no climatic changes can yet be attributed to the CO_2 increase. However, if the concentration continues to increase, a point will be reached when a climatic change can be attributed to the CO_2 increase. Several studies have suggested that a doubling of the CO_2 concentration from its preindustrial value will occur within the next 100 years. Other work shows that doubling the atmospheric CO_2 concentration would lead to a global average surface temperature increase of 1.5–4.5 K with associated (but presently unknown) regional changes of climate.

However, it must be borne in mind that there is a large amount of uncertainty about the CO_2 issue, even though significant advances in knowledge have been made during the last few years. There is uncertainty about the future rates of increase of atmospheric CO_2 because the natural sources and sinks of CO_2 are not completely understood and because the future rates of addition due to fossil-fuel combustion and human interference with the terrestrial biota and soils are not known. In particular, it has recently been argued that, in order for the atmospheric CO_2 level to become threatening, large-scale exploitation of coal and a great increase in the international coal trade would be required. There is uncertainty about the climatic response to CO_2 increase, particularly the regional climatic sensitivity and the *transient* (as opposed to equilibrium) response to forcing. Given all of these uncertainties, it is certainly not justified to call for an immediate stop to the use of fossil fuels. On the other hand, since CO_2 is seen as the largest potential climatic threat at present, it would be prudent to keep energy strategies flexible so that the use of fossil fuels could be controlled at a later date if necessary. At the same time, any adoption of energy-saving technology that would reduce the demand for fossil fuels would reduce the input of CO_2 to the atmosphere and postpone the date of a significant CO_2 increase.

The effect of waste-heat release on global climate will almost certainly not be significant for the scale of energy use considered likely during the next 100 years. On a local and regional scale, however, changes could occur, particularly in cloudiness and precipitation, as a result of waste heat addition. The atmospheric circulation appears to be sensitive to heat inputs of the order of hundreds of TWyr/yr, which are not likely in the next 100 years.

The effect of particles on climate is extremely complex and much improvement of climate models will be required before it can be assessed reliably. Particles in the lower atmosphere can cause changes in horizontal and vertical temperature gradients, cloud amount, and cloud reflectivity and thereby have a number of interacting and potentially significant effects. Gases other than CO_2 that are being added to the atmosphere as a result of mankind's activities mostly have a greenhouse effect similar to that of CO_2 and could have a significant additive effect.

The effect of renewable energy sources (solar, wind, hydropower, ocean thermal, etc.) in the next 100 years is likely to be negligible. This is because these systems will not be deployed on a large enough scale in this time period and because a wide variety of small-scale decentralized systems is likely to be installed. The large-scale deployment of these systems would most probably have climatic effects, at least on the local and regional scales, but they are difficult to evaluate because the systems themselves are often not well defined.

The effect of climate on energy systems involves many interactions. Large amounts of climate data are required to assess the resources for renewable energy systems; when these data are not available there is a requirement for models to interpolate or reconstruct relevant information. It is also clear that climate and weather prediction can be valuable inputs to forecasting energy demand. For the study of the effects of climate on energy supply and demand an adequate theory of the physical basis of climate and climatic change and an ability to predict future climate variability are required.

With regard to the present state of knowledge about the interactions between energy systems and climate, it is clear that there is much uncertainty and the tools for studying the climate system have many shortcomings. If the climate were to change as a result of man's use of energy, present studies suggest that the most likely political response would be adaptation, because it does not require investment now to prevent or compensate for changes in the future. If mankind is going to adapt to man-made climatic changes, one may ask whether it is necessary now to put in a large effort to remove uncertainties about potential changes. The answer must be yes! An improved understanding of the potential consequences of human activities would mean that the world would be able to cope better with future climatic changes and might be warned in time of potentially irreversible and undesirable changes. An improved understanding of the effects of climate would help to reduce international tensions in the event of a man-induced climatic change; international cooperation will be required to achieve this improved knowledge. The present uncertainty about the effects of energy systems on climate suggests that energy strategies should be kept flexible and open, so that any changes that might be deemed necessary can be taken into account. Simultaneously, all efforts at energy conservation will reduce the effect of energy systems on climate.

Appendix

ENERGY UNITS

This brief description is based on the definitions given by McDonald (1981).

The two basic types of energy units are:

— those describing *amounts* of energy;
— those describing *rates* at which energy is supplied, converted, transported, or used.

Amounts of energy have units such as barrels of oil equivalent (boe), tonnes of coal equivalent (tce) or kilowatt-hours of electricity (kWh(e)).

Rates of energy use have units such as million barrels of oil per day (mbd), tonnes of coal equivalent per year (tce/yr), and kilowatt-hours of electricity per year (kWh(e)/yr).

As in the report of the Energy Systems Program Group of IIASA (1981), the unit often used in this book for amounts of energy is the terawatt-year (TWyr).

One terawatt-year (1 TWyr) is equal to 10^{12} Wyr, or equivalently 10^9 kilowatt-years (kWyr), or 10^6 megawatt-years (MWyr), or 10^3 gigawatt-years (GWyr).

The unit generally used in this book for rates of energy supply, conversion, transportation and use is the terawatt-year per year (TWyr/yr).

The *capacity* of energy conversion facilities (generally power stations) is expressed in units of terawatts (TW; $1\,TW = 10^{12}\,W$), gigawatts (GW; $1\,GW = 10^9\,W$), or megawatts (MW; $1\,MW = 10^6\,W$). The capacity of an electricity generating station would be given, for example, as 1 GW(e), where the 'e' stands for 'of electricity'.

REFERENCES

McDonald, A. (1981). Energy in a finite world. Executive summary. Executive Report 4, International Institute for Applied Systems Analysis, Laxenburg, Austria.
Energy Systems Program Group of IIASA (1981). Energy in a finite world: A global systems analysis. Ballinger Publishing Co., Massachusetts.

Subject index